Hedging Your Portfolio

Practical Downside Protection for Long-Term Investors

Joseph Trek

Published by NobleTrex Press

For permissions and other inquiries, write to:

P.O. Box 3132, Framingham, MA 01701, USA

Contents

Introduction

Here is a question that, on the surface, seems completely irrational: Why are you willing to lose money every single year on your house?

If you own a home, you almost certainly purchase homeowners insurance. You pay a premium-let's say $1,500 a year-and at the end of that year, if your house hasn't burned down or flooded, that money is gone. It has evaporated. In pure accounting terms, you have realized a 100% loss on that capital.

Do you pound the table in frustration on New Year's Eve? Do you complain to your spouse that the "house hedging strategy" is underperforming because the kitchen didn't catch fire? Of course not. You accept this small, frictional cost as the price of sleeping soundly. You understand that the premium wasn't a bet that your house *would* burn down; it was a guarantee that your financial life wouldn't be destroyed if it *did*.

Now, look at your investment portfolio. For most people, their retirement accounts represent a pile of wealth arguably as significant as their real estate. Yet, when we talk about protecting this wealth, the logic flips. Investors are told that "insurance" in the stock market is a waste of money. They are told that hedging drags down performance. They are told to simply close their eyes, grit their teeth, and "hold for the long term," even as their net worth gets cut in half during a market crash.

We treat our homes like assets that need protection, but we treat our life savings like chips on a roulette table where the only strategy is to hope the wheel eventually spins our way.

This book is about fixing that disconnect. It is about acknowledging that while optimism is a great life philosophy, it is a terrible risk management strategy.

The Tyranny of the Average

To understand why hedging matters, we have to dismantle the most dangerous phrase in finance: "The stock market returns 10% on average."

Mathematically, this might be true over a century. But you do not live in the average. You live in the specific. You live in the messy, volatile, heart-stopping reality of the path *to* that average. If the market drops 50% this year and rises 50% next year, your average return is 0%. But your money? You lost 25% of it. That is the cruel geometry of compounding losses. A 50% drop requires a 100% gain just to get back to square one. That isn't just a math problem; it's a time problem. If that drop happens five years before you retire, you don't have the time to wait for the "average" to show up and save you.

This is where the traditional advice-"Just buy a diversified portfolio of stocks and bonds"-starts to crack. Diversification is wonderful. It is the only free lunch in finance. But in a true liquidity crisis, like we saw in 2008 or the flash crash of 2020, correlations go to one. Everything falls together. Bonds might cushion the blow, or they might fall alongside stocks if interest rates are rising. Relying solely on diversification is like relying on the airbags in your car; they help, but wouldn't you prefer to have brakes?

Hedging is your braking system. It is the art of intentionally structuring your portfolio so that when the road gets icy, you don't careen off the cliff.

The Toolkit You Were Told Was Too Dangerous

For decades, the financial industry has kept the best protective tools behind a velvet rope. Derivatives-specifically options-are frequently demonized in the financial press as "weapons of mass destruction" or tools for reckless gamblers. And, to be fair, they can be. If you hand a chainsaw to a toddler, you get a disaster. But if you put that chainsaw in the hands of a skilled arborist, you get a healthy forest.

The instruments we will discuss in these pages-puts, collars, futures, and volatility products-are just tools. They are neutral. They take on the intent of the user. A speculator uses options to leverage risk and chase lottery-ticket returns. A hedger uses the exact same options to *delete* risk.

We are going to demystify these tools. You don't need a PhD in stochastic calculus to buy a protective put. You don't need to be a Wall Street trader to understand how a collar works. You just need to understand the mechanics of insurance. We will strip away the jargon and look at the payoffs. We will treat these contracts not as scary financial voodoo, but as customizable insurance policies that you can design to fit your specific fears.

The Cost of Safety

Let's be honest with each other before we go any further: Hedging is not free.

If anyone tries to sell you a strategy that captures 100% of the market's upside with zero risk of loss, run away. They are lying to you. In financial markets, risk and return are two sides of the same coin. If you chop off the risk tail, you usually have to pay for it, either through direct costs (buying options) or opportunity costs (capping your upside).

This book is not about magic. It is about trade-offs. We are going to have a very adult conversation about the "cost of carry." We will look at why protection is often expensive, why the market charges a premium for it, and how to value it correctly.

The question isn't "Is hedging expensive?" The question is "Is the cost of the hedge cheaper than the cost of the ruin it prevents?"

If you pay 1% or 2% of your portfolio a year to ensure you never face a drawdown deeper than 15%, you will likely underperform the S&P 500 during a raging bull market. You need to be okay with that. The goal of hedging is not to beat the market every single year. The goal is to survive the years that wipe everyone else out. It is about compounding your wealth uninterrupted. As the old saying goes: Rule No. 1 is never lose money. Rule No. 2 is never forget Rule No. 1.

Beyond the Math: The Sleep Factor

While we will spend a lot of time on payoff diagrams, Greeks, and volatility surfaces, the most important benefit of hedging isn't mathematical. It's behavioral.

The biggest threat to your portfolio isn't the Federal Reserve, inflation, or a global pandemic. It is the person looking back at you in the mirror. When markets crash, panic sets in. The lizard brain takes over. Investors sell at the bottom because the pain of watching their net worth evaporate becomes physically intolerant.

A well-hedged portfolio solves this. When you know that you have a "floor" beneath you-a mathematical guarantee that you cannot lose more than a set amount-panic is replaced by strategy. You don't sell the low; you might even buy it. The hedge acts as a psychological anchor, keeping you rational when the rest of the world is losing its mind.

How to Use This Book

We are going to build this knowledge from the ground up, moving from the philosophical to the highly tactical.

We will start by looking at the **Case for Protection**. We need to agree on what we are actually trying to solve. Are we worried about a 10% correction or a 40% crash? Are we protecting a retirement nest egg that we need to tap in two years, or a multi-generational trust? The answers change the strategy.

Then, we will open the **Hedger's Toolkit**. We will get our hands dirty with the mechanics of options. If you've never traded an option before, don't worry. We will walk through how they work, how they are priced, and why they act the way they do. We will learn to read "payoff diagrams"-the sheet music of hedging-so you can visualize exactly what your portfolio will do in any market scenario.

Once we know the tools, we will look at **The Price of Insurance**. Understanding *implied volatility* is the superpower of the professional hedger. You wouldn't buy car insurance if the premium cost more than the car, yet investors often buy "crash protection" when it is mathematically overpriced. We will learn how to spot value and how to avoid being ripped off by the market's fear pricing.

From there, we move to **Core Structures**. This is the meat of the book. We will detail specific strategies: the protective put, the collar, put spreads, and how to use index proxies to hedge stock portfolios. We won't just look at the theory; we will look at the maintenance. A hedge is not a "set it and forget it" trade. It is a living thing that decays, changes, and needs to be rolled. We will discuss the **Strategy Design and Maintenance** required to keep your defenses up without bleeding money.

Finally, we will tackle the **Operational Reality** and the **Human Element**. Theory is great, but execution is everything. What happens when liquidity dries up? What about taxes? How do you explain this to your spouse or your investment committee? We will provide a governance framework to help you stick to the plan when emotions run high.

A New Way of Thinking

Reading this book requires a shift in mindset. You are moving from being a passive passenger in the market to an active risk manager.

Most investors operate with a binary view: they are either "bullish" (all in) or "bearish" (all cash). This is a stressful, inefficient way to live. The hedger exists in the gray area. The hedger can be long the market but skeptical of the risks. The hedger can participate in the upside while carrying a parachute.

This approach is not for everyone. It requires effort. It requires learning a new vocabulary. It requires the discipline to pay for protection when the sun is shining and the market is hitting all-time highs. But for the long-term investor who understands the destructive power of drawdowns, it is the only way to truly secure financial freedom.

The markets are uncertain. They are prone to shocks, manias, and panics. We cannot control what the market does. But we can absolutely control how our portfolio reacts to it.

Let's get to work.

Chapter 1

The Case for Protection: Foundations of Risk Control

If you lose 50% of your portfolio, you don't need a 50% gain to get back to even-you need 100%. That math, cruel and unforgiving, is the heartbeat of this chapter. We move beyond standard diversification advice to explore why traditional safety nets often tear when you need them most, and how true risk control is less about predicting the future and more about ensuring you survive it.

1.1 The `Why' Before the `How': Matching Defense to Your Goals

A friend of mine once bought a used sports car that had been modified for track days. It looked incredible. It sounded like speed. It also scraped on speed bumps, hated potholes, and made every grocery run feel like a minor act of courage. The car wasn't "bad." It was simply built for a different job.

Portfolios are like that. Some are built to win races. Some are built to get the kids to school in the rain with a trunk full of groceries. The mistake is not choosing one or the other—the mistake is pretending they're the same vehicle, then acting surprised when the suspension snaps.

Downside protection starts with a question that sounds almost too basic: *What is this money for?* Not "what do I want to earn?" but "what must not go wrong?" A portfolio meant to fund a vacation home in 2045 can survive a few ugly years. A portfolio meant to cover tuition in 18 months cannot. The second portfolio doesn't have the luxury of being "right eventually." It needs to be solvent *on time*.

That idea—*on time*—is the real heartbeat of hedging. The stock market has recovered from every historical crash in the broad sense, but individual investors don't live inside "the broad sense." They live inside calendars, bills, and bodies that get older. The most painful market losses aren't the ones that look dramatic on a chart; they're the ones that collide with your life at the wrong moment.

Risk tolerance is a personality quiz; risk capacity is a balance sheet

"Risk tolerance" gets talked about as if it's an emotional trait, like being introverted or extroverted. And sure—some people can watch their net worth swing by six figures and sleep like a stone. Others feel physically ill when an index is down 3%. But the deeper, more practical concept is *risk capacity*: your *financial* ability to endure losses without your plans breaking.

Risk tolerance is how much volatility you *think* you can stomach. Risk capacity is how much volatility your *life* can absorb.

The two often disagree, and when they do, reality wins. A nervous but financially secure investor can choose to take less risk than they can technically afford; they may give up some upside, but they won't blow up their future. A confident investor with thin margins—high leverage, big near-term obligations, little cash cushion—can feel brave right up until the market sends the invoice.

A good way to feel this difference is to imagine a simple stress test: *If my portfolio dropped 30% by next month, what would I be forced to do?* Not what would you *want* to do, but what would you be *forced* to do. Sell to pay the tax bill? Sell to make a down payment? Sell because the mortgage feels unsafe? Sell because you'd lose your job in the same recession that caused the drop?

If the honest answer includes the word "sell," you've discovered something important: you're not just exposed to volatility—you're exposed to *timing*. That's when protection stops being an academic concept and starts resembling a seatbelt.

Time horizon isn't a number; it's the date of your next unavoidable withdrawal

Investors love to say, "I'm long-term." It sounds disciplined. It sounds professional. It can also be meaningless.

Your real time horizon is the first time you'll need to pull meaningful cash out of the portfolio for a non-negotiable reason. That could be retirement spending. It could be a home purchase. It could be tuition. It could be a business buy-in. It could be the moment you plan to become a single-income household. It could be medical expenses you don't fully control.

A portfolio can be "for retirement in 25 years" and still have a 2-year time horizon if you also intend to use it for a down payment. The label on the jar doesn't matter; the withdrawal date does.

This is why downside protection is sometimes the most rational choice for the investor who seems, on paper, the most conservative. If you have a big liability coming due in a short window, volatility isn't "risk" in the abstract—it's a threat to a specific plan. In that context, hedging isn't a way to get fancy; it's a way to avoid becoming a forced seller at the worst possible time.

On the other hand, if you truly have decades before you need the money, the same hedge can behave like a slow leak. It may reduce the very thing you need most—long-term compounding—while protecting you from a danger you can already survive: temporary drawdowns.

The hard truth is that many investors buy protection because they dislike seeing losses, not because losses are actually lethal to their plan. That can be an expensive form of comfort.

The race car versus the family sedan: human capital changes everything

There's a hidden asset on your personal balance sheet that rarely shows up in portfolio conversations: *human capital*. That's the present value of your future earnings—your ability to work, earn, and rebuild.

A 25-year-old engineer with a stable job has a massive "bond-like" asset in the background: future paychecks. Even if their portfolio gets cut in half, their life doesn't necessarily change. They can keep contributing, buying shares at lower prices, and letting time do its slow, brutal magic in reverse. Volatility can actually help them if it brings lower prices early in the savings journey.

A 70-year-old retiree has almost the opposite situation. Their portfolio *is* their paycheck. Their human capital is mostly spent. A severe drawdown doesn't just sting—it reduces the fuel available for everything that comes next. They can't easily "wait it out" without changing spending, because waiting still costs money.

Protection is often most valuable when human capital is low and financial obligations are high. That's not a moral statement; it's geometry. When you can't replenish the bucket with labor, you have to worry more about the holes.

Here's the twist that surprises people: the highest *emotional* appetite for risk sometimes shows up precisely when risk capacity is lowest. Retirees who've invested for decades can feel battle-hardened, even invincible. They've seen crashes before. They've lived through headlines. But sequence risk—the danger of a bad run early in withdrawals—doesn't care how tough you feel. It cares how many shares you're forced to sell when prices are depressed.

If you're still earning and contributing, a downturn can be a sale. If you're withdrawing, a downturn can be a trap.

Liabilities: the ``short position" hiding in your life

Most people understand a short position as a bet against something. But in personal finance, you can be short without ever touching a trading account. A liability is a future obligation that behaves like a negative asset. It is, in a sense, a claim on your future wealth.

Some liabilities are obvious: mortgages, tuition, planned purchases, tax bills. Others are sneakier: the desire to maintain a certain standard of living, the commitment to help aging parents, the intention to leave a bequest, the promise you made (explicitly or implicitly) that you won't become financially dependent on your children.

A portfolio without reference to liabilities is like measuring a bridge's strength without knowing the weight of the trucks that will cross it.

Liabilities also come with deadlines. A mortgage payment is due monthly whether markets are kind or cruel. Tuition bills arrive on schedule. Health events can show up without notice and demand immediate liquidity. The market is optional; the liability often is not.

This is one reason "100% stocks forever" is not a universal truth, even for investors who believe deeply in equities. A liability turns volatility into something more dangerous than temporary mark-to-market pain. It turns volatility into a potential *mismatch*—your assets may be down precisely when your obligations are due.

Once you see liabilities as the real enemy, hedging becomes less about prediction and more about synchronization. You are trying to align the behavior of your assets with the timing of your needs.

The two kinds of pain: paper cuts versus broken bones

Not all losses are created equal. A 15% drawdown feels bad, but it's often survivable. A 50% drawdown can permanently alter what's possible, especially if it arrives near a spending event or a career disruption.

The practical question is not "How do I avoid losses?" That's fantasy. The useful question is "Which losses would force me to do something irreversible?"

Irreversible actions include:

- Selling at depressed prices to fund spending.
- Moving to a lower standard of living permanently because the portfolio can't support the old one.
- Delaying retirement by years because the portfolio's new baseline is too low.
- Taking on leverage at the wrong time to "catch up."
- Concentrating risk to recover faster—turning a bad drawdown into a catastrophic one.

Notice what's not on that list: "Feeling bad." Emotional discomfort matters—investors do make terrible decisions under stress—but it's still different from structural fragility. Protection is easiest to justify when it prevents broken-bone outcomes, not when it merely reduces paper cuts.

This is where many investors get the hedging decision backward. They start with a tool ("I should buy puts" or "I should hold gold" or "I should go to cash") and then try to invent a reason. The reason has to come first. Without it, you're just paying for complexity.

A simple way to map your ``must not fail" money

Most households have at least three categories of money, even if they don't label them:

1. *Soon money*: spending in the next 0–3 years (emergency fund, near-term purchases, taxes, tuition deposits).
2. *Schedule money*: goals with a date you care about (down payment in 4 years, retirement in 7 years, a sabbatical in 2 years, business investment next year).
3. *Forever money*: money you won't need for a long time, and where the main enemy is inflation rather than volatility.

Only the third bucket has the luxury of being "unhedged by default." The first two buckets tend to be where protection earns its keep—not necessarily with options, but with a conscious reduction in drawdown risk. That might mean cash, short-term Treasuries, a bond ladder, a less aggressive allocation, or in some cases explicit hedges.

"Forever money" can still be hedged, especially if an investor has unique constraints or cannot emotionally stay invested, but the burden of proof is higher. You need a clear reason why reducing volatility today improves the probability of reaching the long-term goal *more* than it reduces it by lowering expected returns.

The most common reason is behavioral: a hedge that keeps you invested during panics can be worth far more than its explicit cost. But that only works if the hedge is designed to actually pay off when you're tempted to abandon the plan—not just in mild corrections where everyone feels clever.

When hedging is a necessity (and when it's a drag)

Hedging becomes closer to necessity when:

- You have a large, near-term liability or spending need that you cannot postpone.
- Your job or business income is highly correlated with the stock market (tech employees with heavy equity comp, cyclical industries, financial sector exposure).
- Your portfolio must support withdrawals soon, and you cannot meaningfully cut spending.
- You are using leverage (margin, large mortgages relative to liquid assets, or concentrated positions that can trigger forced sales).
- A drawdown would cause an irreversible life decision you deeply want to avoid.

Hedging tends to be a drag when:

- You have a long horizon, stable income, and the ability to keep buying through downturns.

- Your "risk" is mostly emotional discomfort rather than financial fragility.
- You buy protection continuously without a clear definition of what outcome you're insuring against.
- The hedge is sized so small that it can't realistically change your behavior or your plan in a real crash—meaning you pay for peace of mind but don't actually receive it.

One of the least discussed traps is the "decorative hedge": a strategy that looks sophisticated but is too small, too short-dated, or too inconsistent to matter when it counts. It can create the illusion of prudence while leaving the underlying portfolio exposed. Worse, it can encourage taking more risk elsewhere—like wearing a helmet and then deciding seatbelts are optional.

A real hedge has to be tied to a real scenario. *What exactly am I trying to survive?* A 20% correction? A 35% drawdown? A 2008-style crisis? A lost job plus a bear market? The answer dictates the tool, the sizing, and whether the ongoing cost is justified.

The hidden correlation: your paycheck and your portfolio

Diversification is often explained as "don't put all your eggs in one basket." But most people already have a giant basket: their career. If your income depends on the same economic engine that drives your investments, your life is more concentrated than your brokerage statement suggests.

Think of a venture-backed startup employee who holds company stock and invests heavily in growth equities. Their paycheck, their equity compensation, and their portfolio can all be hit by the same shock: tightening financial conditions. That's not diversification; it's stacking dominoes.

Or consider a real estate professional whose income and investment portfolio both depend on the health of property markets and credit availability. A housing downturn doesn't just reduce the value of their assets; it can simultaneously reduce deal flow, commissions, and financing options.

In these cases, hedging can be less about portfolio theory and more about household survival. You're protecting against a world where the same event hurts your earnings *and* your invested wealth. That's the scenario that forces rushed decisions.

If you can't diversify your paycheck, you may need to diversify your portfolio more aggressively, or consider targeted hedges that pay off in recessions. The point is not to become pessimistic—it's to avoid being *fragile.*

A practical definition of "enough protection"

Investors often ask, "How much should I hedge?" The better question is, "How much loss can I absorb *without changing the plan*?"

That number is different for everyone. For one household, a 25% drawdown means delaying a kitchen remodel. For another, it means downsizing a home, abandoning travel plans, or returning to work. Those are different stakes, and they deserve different defenses.

One concrete exercise is to pick a "plan survival" threshold:

> *If my portfolio fell by X%, my plan would still work without requiring painful, permanent changes.*

Then choose X honestly. It is not a badge of honor to pick a bigger number than your life can actually handle. Pride is not a risk-management tool.

Once you know your threshold, protection becomes easier to evaluate. If a hedge reduces the probability of breaching that threshold, it's doing real work. If it merely smooths returns slightly, it may feel good, but it might not be worth the cost.

This is also where investors discover that "hedging" can be as simple as holding a cash reserve large enough to avoid selling in a downturn. Cash is not glamorous. It also does something options can't always do: it buys time. In a crisis, time is often the most valuable asset in the world.

The emotional side: hedging as a commitment device

Even disciplined investors can be surprised by how a true bear market feels. Screens turn red day after day. Confidence becomes exhausting. Friends who never cared about markets start asking if you're worried. The news stops being informational and becomes theatrical. You begin to question whether the world has changed permanently.

In that environment, the best-designed asset allocation in the world can fail for a very human reason: you stop following it.

This is the most underappreciated argument for selective hedging. A hedge can act like a commitment device—a prepaid decision that helps you hold steady when your instincts scream to do something dramatic.

But only if the hedge is credible. If it's too small to matter, it won't change your behavior. If it's too complicated, you might not trust it when volatility spikes. If it requires perfect timing, it can become another source of stress.

The ideal behavioral hedge is boring in design and comforting in function: you understand it, you expect it to cost something, and you know the conditions under which it should help. It doesn't need to make you money. It needs to keep you from making a life-altering mistake.

Actionable takeaway: write your ``risk capacity memo'' before you shop for hedges

Before considering any protective strategy—options, tactical shifts, alternative assets, anything—write a one-page memo to yourself with three sentences:

1. *This money is for:* (name the goal and the date it must work).
2. *My plan breaks if:* (define the drawdown or scenario that would force irreversible changes).
3. *I can tolerate:* (a realistic maximum loss without changing contributions, spending, or retirement timing).

If you can't fill those in, you don't need a hedge yet—you need clarity. If you can fill them in, you've just done the hardest part of downside protection: you've turned "fear of losses" into a specific engineering problem. Only then does it make sense to ask *how* to build the defense.

1.2 When Diversification Disappears

The first time most investors learn what "diversification" really means is not in a classroom—it's in a bad week. They open their account expecting at least *something* to be green. Maybe international stocks will save the day. Maybe bonds will do their dependable, boring job. Maybe gold will play hero. Instead, everything looks bruised. The portfolio feels less like a carefully balanced meal and more like a single, oversized bet that just went wrong.

That moment creates a particular kind of anger. Not the hot anger of losing money—more like the cold anger of being misled. *I did the responsible thing. I diversified. Why am I still bleeding?*

The honest answer is that asset allocation is the first line of defense, not the last. It's a remarkably good tool for most market weather. It just isn't built to handle every kind of storm. Diversification can reduce risk in the way shock absorbers reduce bumps: most of the time, it makes the ride tolerable. But if you hit a pothole big enough to crack the rim, you discover the limits of suspension.

Diversification is not ``own more stuff." It's ``own different *behaviors*."

Diversification is often taught as a shopping list: U.S. stocks, international stocks, bonds, maybe a slice of real estate. But the deeper idea is behavioral. A diversified portfolio is one where different parts tend to react differently to the same headline.

In plain language, you're trying to avoid a portfolio where every holding has the same bad day for the same reason.

That's why "more holdings" is not automatically "more diversified." Owning 200 technology stocks is a concentrated bet dressed up as variety. Owning a total stock market index plus a small-cap index plus a growth ETF can still be one big equity engine with three different paint jobs. The portfolio may look complex while still being exposed to the same core risk: a broad repricing of equities.

A good asset allocation usually starts with a simple recognition: stocks are phenomenal long-term growth engines, but they are emotionally violent. Bonds tend to offer lower long-term returns, but they often provide ballast—*when* they do, they can be the piece you rebalance from, the dry powder you use to buy stocks when they're on sale, the part of the portfolio that keeps you from panicking.

That "when" matters more than people like to admit.

The quiet promise inside a classic stock/bond mix

The traditional diversified portfolio—some stocks, some bonds—carries an implicit promise: when stocks fall, bonds will either rise or at least hold steady, giving you something to lean on.

Sometimes this works beautifully. You get a selloff in equities, yields fall, bond prices rise, and suddenly the boring part of the portfolio is doing something heroic. Rebalancing becomes a rational act instead of a leap of faith. You sell a little of what went up (bonds) to buy what went down (stocks), and you feel like an adult.

But that promise is not a law of nature. It's a pattern that depends on the type of shock, the inflation environment, and where interest rates started.

In an inflation scare, both stocks and bonds can struggle together. Stocks fall because future earnings are discounted at higher rates and margins get squeezed. Bonds fall because their fixed payments are suddenly less attractive relative to new, higher yields. In that world, the "ballast" becomes another weight.

If you've never lived through a period where bonds didn't feel like safety, it can be disorienting. It's like discovering that your umbrella was only tested indoors.

Correlation: the word you ignore until it starts shouting

Correlation is simply a measure of how two things move together. A correlation near −1 means they tend to move in opposite directions. Near 0 means their moves are mostly unrelated. Near +1 means they move together.

In calm markets, correlations can be pleasantly low. Different countries have different economic cycles. Different sectors have different fortunes. Bonds often have their own rhythm. Diversification feels like a smart, almost magical hack: you can keep a meaningful allocation to growth assets while smoothing the ride.

In crisis markets, correlation has a nasty habit of rising. It's not that every asset becomes identical, but the portfolio starts responding to a single dominant force: *risk-off*. People stop caring about nuance. They sell what they can, not just what they want to. Prices move less on thoughtful valuation and more on urgent balance-sheet management.

That's why, in the ugliest moments, investors joke that "the only thing that goes up is correlation." The joke lands because it feels true when you need diversification most.

What a ``liquidity crisis" looks like from the inside

A liquidity crisis is not just a period when prices fall. It's a period when *the ability to trade at a fair price* disappears. Spreads widen. Market depth thins. Funds and institutions face redemptions. Leverage gets squeezed. Margin calls start acting like gravity.

When forced selling enters the room, diversification can weaken because selling becomes indiscriminate. If you manage a portfolio and you need cash *today*, you don't necessarily sell the asset with the weakest long-term prospects. You sell the asset that you can sell quickly, in size, without breaking the market too badly. That often means selling liquid "good" assets alongside the risky ones.

In that environment, "diversifiers" can get dragged down simply because they're easy to sell.

This is also why investors sometimes watch gold, quality stocks, or even certain bond sectors dip at the same time as everything else, then recover later. The initial phase is about liquidity and survival, not about thoughtful asset pricing. Later, after the forced selling passes, fundamentals start mattering again—and the portfolio's diversifiers may finally do their job. But that lag can be brutal if you needed protection *immediately*.

The two flavors of diversification failure

Diversification doesn't fail in only one way. It fails in at least two distinct flavors, and confusing them leads to the wrong fixes.

Flavor 1: "Everything risky fell together." This is the classic correlation spike. Equities across regions fall. Credit spreads blow out. Real estate wobbles. Even "defensive" equity sectors get hit because investors aren't trying to pick winners—they're trying to cut risk.

In this scenario, diversification within equities (U.S. vs. international, growth vs. value, sector mixes) can help on the margins, but the dominant driver is equity beta: the market's overall risk appetite.

Flavor 2: "The diversifier fell too." This is the more psychologically damaging one. The asset you were counting on—often bonds—also declines. Now you don't just have drawdown pain; you have worldview pain. The portfolio didn't just lose money; it lost its *design logic*.

This second flavor is often tied to inflation dynamics and interest-rate shocks. It's not that bonds can't diversify equities. It's that they don't diversify equities *in every regime*.

The right response depends on which failure mode you're facing. If the issue is "equities fell together," you may need a true crash hedge. If the issue is "bonds didn't ballast," you may need to rethink duration exposure, inflation sensitivity, or what you consider a diversifier in the first place.

Why international diversification can disappoint right on schedule

International diversification is a perfect example of a concept that is true in theory and inconsistent in practice.

Over long periods, different countries do experience different cycles. Currency moves can also provide diversification. But in global crises, markets often behave like one big room with one big exit. Multinational companies are exposed to global supply chains, global credit conditions, and global investor sentiment. In a worldwide selloff, "international" can feel less like a hedge and more like the same pain with a different time zone.

There's also a structural reality: global equity markets are linked through capital flows. When large institutions reduce risk, they often reduce risk everywhere. The selling pressure itself becomes a channel of contagion.

That doesn't mean international exposure is useless—far from it. It can reduce concentration in any single economy and can improve long-run resilience. The point is narrower and more practical: if your plan assumes international stocks will reliably zig when U.S. stocks zag during panics, you may be leaning on a cane that sometimes turns into a noodle.

``But I owned defensive stocks." Defensive *relative* to what?

In a mild correction, "defensive" stocks can feel like a safe corner: consumer staples, utilities, healthcare. They have steadier cash flows. People still buy soap and electricity in recessions. Great.

In a true crisis, two things happen.

First, correlations inside the equity market can rise. Defensive sectors may fall less, but they still fall, because they are still equities. If investors are reducing equity exposure broadly—whether due to risk limits, margin pressure, or fear—defensive names get sold too. They're just less likely to be the first to get crushed.

Second, valuation matters. If defensive sectors were priced expensively going into the crisis (because everyone had the same "safety" idea), they can be surprisingly fragile. Paying a premium for perceived safety is a common late-cycle habit. When the market resets, expensive "safe" assets can drop harder than expected.

Defensive equities can be useful. They are not the same thing as protection. They are speed limiters, not airbags.

The nuance: diversification doesn't ``disappear" equally for every goal

Here's the subtle point that saves you from cynical fatalism: diversification doesn't have to be perfect to be valuable. It just has to be valuable *for your objective.*

If your objective is to maximize long-term wealth over decades, then even imperfect diversification can be a meaningful improvement. Shaving a little volatility, reducing the frequency of deep drawdowns, and keeping you invested can compound into a huge difference.

If your objective is to survive a short-window liability—say you need a down payment in 18 months—then "imperfect" diversification may not be good enough. A 25% drawdown at the wrong moment is not a statistical inconvenience; it's a plan failure.

The same portfolio can be "diversified" for one person and dangerously fragile for another because the goal changes what "risk" means. A retiree funding withdrawals and a young saver funding contributions live in different mathematical universes, even if they hold the same funds.

So when someone says "diversification failed," the right follow-up question is: *failed at what?* Failed to avoid a drawdown? Failed to provide rebalance fuel? Failed to preserve spending power? Failed to keep you calm enough not to sell?

A portfolio can fail one of those and still succeed at the others.

A more realistic mental model: diversification buys time, not invincibility

A well-built allocation often buys you something precious in bad markets: time. It may reduce the pace of losses. It may give you something to rebalance from. It may dampen volatility enough to keep you from panic-selling.

But in a severe systemic event, time may be all it buys. If you're depending on diversification to deliver *positive* returns during a crash, you're asking too much of it.

That's not a reason to abandon asset allocation. It's a reason to treat it like the foundation, not the roof. You still want the foundation. You just don't want to confuse it with a bunker.

What actually helps when correlations rise

When correlations converge, you need assets or strategies that are designed to behave differently *in that regime*. Broadly, there are a few categories of "different behavior" that matter during panics:

- *Cash and near-cash*: not exciting, but immediately spendable and not forced to take market risk.
- *High-quality government bonds (sometimes)*: especially when the shock is deflationary or growth-slowing and rates can fall.
- *Explicit convexity*: strategies that can gain more as losses deepen (often options-based, like protective puts).
- *Structural income floors*: pensions, annuities, or other guaranteed income sources that reduce the need to sell assets in a downturn.

Notice how different these are from "more equities, but different." In crisis regimes, the portfolio often needs genuinely different engines, not just different seat covers.

This is also where the earlier goal-matching work pays off. If your plan breaks at a certain drawdown threshold, you may need a tool that is designed to respond specifically to deep drawdowns. Traditional diversification can reduce the odds of getting there, but it may not cap the damage once you do.

The uncomfortable truth: diversification is visible in hindsight, protection is visible in advance

Diversification feels good because it is socially approved and historically defensible. You can point to charts. You can cite long-term averages. It's the financial equivalent of eating vegetables.

Protection—real protection—often feels psychologically harder because it admits vulnerability. It says, "There are outcomes I can't afford, even if they're unlikely." It also forces you to accept ongoing costs: lower expected return, premium spend, or the embarrassment of watching the hedge lose money in good years.

That's why many investors stop at diversification. It's a clean story. It's a respectable story. And for many goals, it's enough.

The moment diversification becomes insufficient is not when the market gets scary. It's when the market's scary behavior intersects with your personal deadlines, liabilities, and constraints—when a drawdown becomes not just painful but *consequential.*

Actionable takeaway: stress-test your allocation for a ``one-exit" day

Once a year—seriously, put it on the calendar—run a simple, brutally honest scenario:

> *Assume a fast, ugly drawdown where most risky assets fall together. What, specifically, in my portfolio would I*

> *sell to raise one year of spending (or to meet my next major liability) without locking in catastrophic losses?*

If you don't have a clear answer—if the plan is basically "hope stocks bounce back soon"—then your diversification may be a fair-weather friend. That doesn't mean you need to overhaul everything. It means you should consider adding something that creates *decision-making oxygen* in a panic: a bigger cash buffer, better liability matching, less reliance on a single diversifier, or a defined-cost hedge sized to matter.

The point isn't to build a portfolio that never bleeds. It's to build one that doesn't force you into the worst decision of your investing life on the worst day of the market.

1.3 The Cruel Math of Big Losses

There"s a particular moment in every real bear market when the conversation changes tone. Early on, people say things like "healthy correction" and "buying opportunity." Later—usually after the losses stop being tidy—people start speaking in confessions. Plans get postponed. Spending gets quietly trimmed. The word "retirement" gets used less like a date and more like a weather forecast.

What changed wasn't just sentiment. The math changed.

Small losses are annoying. Big losses are transformative. The market doesn't need to "end" to permanently alter your financial trajectory; it only needs to push you down a hill that is steeper to climb than it looks.

Losses and gains are not mirror images

A gain and a loss of the same percentage do not cancel each other out. That sounds like a trick question, but it's the kind of trick that can cost years.

If you have $100 and you lose 50%, you're at $50. A 50% gain from there gets you to $75—not back to $100. You're still short. The loss shrank the base you're compounding from, so every future percentage return applies to a smaller number.

To get back to $100 after dropping to $50, you need a 100% gain. That's the asymmetry. Losses dig a hole; gains fill it more slowly than you expect because the shovel is smaller.

Investors sometimes react to this by trying to "earn it back" quickly. That instinct is deeply human and often deeply destructive. The market's cruelest trick is that after a large loss, the desire to take more risk feels like a solution when it's actually gasoline.

The Recovery Hill: why the climb gets steep fast

A 10% loss needs an 11.1% gain to recover. A 20% loss needs a 25% gain. A 30% loss needs about a 42.9% gain. A 40% loss needs about a 66.7% gain. A 50% loss needs a 100% gain.

Those numbers are not opinions. They are arithmetic.

And they have an emotional implication that's easy to miss: the deeper the drawdown, the more your recovery depends on either (a) time, or (b) taking additional risk, or (c) adding new money. If you don't have (c), and you don't have patience for (a), people often reach for (b). That's when portfolios start drifting from "investing" into "trying to get back to normal."

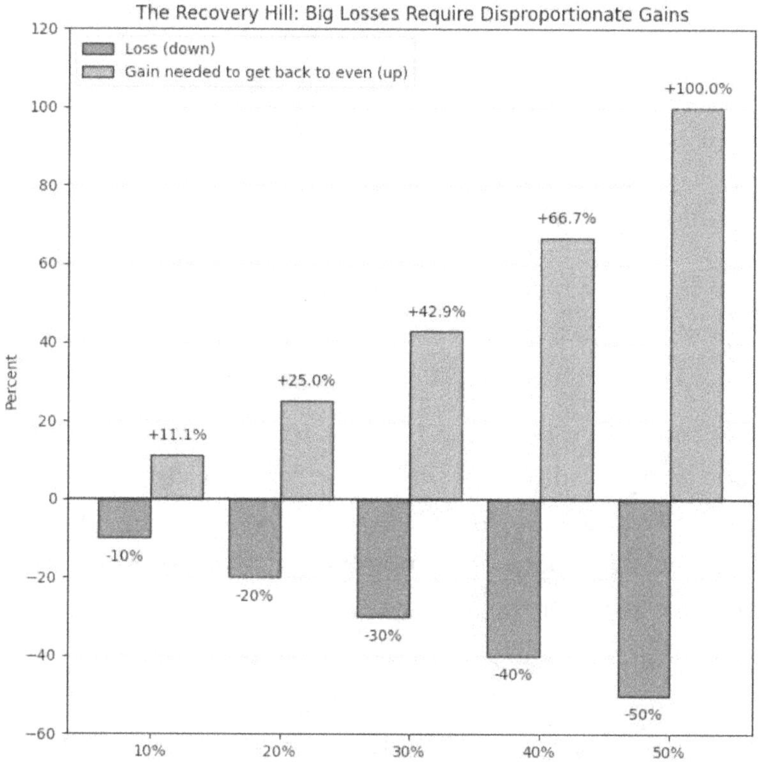

Look at the rightmost pair of bars—the 50% loss and the 100% recovery requirement—and notice what it implies about *time*. A diversified stock portfolio might reasonably earn, say, high single digits over long periods, but it does not hand out 100% gains on demand. A 100% gain is the kind of thing that can happen, but it can also take many years, especially if the recovery path includes volatility and false starts.

That's why deep drawdowns don't just hurt; they *consume calendar*. And your calendar is not an abstract axis on a chart. It contains tuition dates, retirement dates, health events, family needs, and the plain fact that you do not get infinite do-overs.

Drawdown is the silent partner in every compounding story

Compounding is often sold as a friendly miracle: money making money making money. But compounding has a rival that is just as powerful and much less celebrated: drawdown.

A portfolio that compounds at 8% for decades can build extraordinary wealth. A portfolio that compounds at 8% but occasionally suffers a severe drawdown can end up in a very different place, because compounding works on the amount you have left after the damage.

It's tempting to say, "I'll just hold through it." Sometimes you can. Sometimes you can't. The phrase "hold through it" assumes you have no cash needs, no job risk, no psychological breaking point, and no leverage. It assumes your life politely pauses while the market does its thing.

Even if you do hold, deep drawdowns still matter because they change the *distribution* of outcomes. They increase the chance that you'll be behind schedule at precisely the wrong moment. They increase the temptation to abandon the plan. They increase the odds that you'll make a "small" permanent shift—like staying underinvested for a few years after the crash—that costs far more than the crash itself.

The biggest losses often don't do their harm on the day they happen. They do their harm in the decisions they provoke afterward.

Volatility drag: the leak in the wealth bucket

"Volatility drag" sounds like something a quant would say while staring at a spreadsheet. In real life, it's the reason two portfolios with the same average return can end up with different results.

Here's the intuition without the jargon: when returns bounce around, losses hit harder than gains help, because losses shrink the base. The more violently a portfolio swings, the more often it has to recover from dips before it can make net progress.

A simple illustration makes this visceral. Imagine a portfolio that goes down 20% one year and up 20% the next. The average return is 0% (because -20% and +20% average to 0%). But the portfolio is not flat. $100 becomes $80 after the first year, then $80 becomes $96 after the second. You lost 4% despite a "0% average."

That missing $4 is volatility drag—money that leaked out through the math of bouncing around.

Real markets are messier than a two-year toy example, but the principle scales. Higher volatility means more drag. Deep drawdowns are the extreme form of the same problem: a very large "down" move that demands a disproportionate "up" move to get back.

This is why investors who focus only on average returns can misread risk. Averages don't tell you how rough the path was, and the path can be the difference between reaching a goal and missing it.

The psychological tax: big losses make you pay twice

The first tax is the financial one: you need a larger gain to recover. The second tax is psychological: big losses make it harder to stay invested long enough to earn that gain.

After a mild decline, people talk about buying more. After a severe decline, people talk about "waiting for things to settle down." That phrase sounds prudent, but it often means "I will re-enter after prices are higher and fear is lower," which is a polite way of locking in opportunity cost.

This is not a character flaw. It's a wiring issue. Humans are built to avoid pain, and markets deliver pain in a format that feels personal. Your brain experiences a drawdown as a threat that demands action. The market often rewards the opposite: patience, rebalancing, and doing nothing dramatic.

Protection—whether through conservative allocation, cash reserves, or explicit hedges—can reduce the size of the psychological tax by reducing the size of the drawdown. The goal isn't to become numb. The goal is to keep losses in a range where you can still behave like the person you planned to be.

Why ``I'll just wait for it to come back" is not always available

The phrase "it always comes back" is true for broad markets in a historical sense, but it can fail you in three very practical ways.

First, *you may not own the whole market*. Concentrated positions can fail permanently. Individual companies do not "always come back." Sectors can underperform for long stretches. Countries can have lost decades. "Always comes back" is an index investor's statement, not a universal guarantee.

Second, *your life may not wait*. If withdrawals begin, if a down payment is due, if a job loss coincides with a recession, you can be forced to sell before the recovery arrives. The market's long-term resilience doesn't help you if your cash flow calendar demands liquidity now.

Third, *your future self may not cooperate*. Many investors plan to hold through downturns, then discover that their fear threshold is lower than expected. That's not hypocrisy; it's new information. Severe drawdowns reveal the difference between theoretical discipline and practiced discipline.

Deep losses are dangerous partly because they compress all three of these risks into the same time window: market stress, economic stress, and personal stress. That's when diversification's limits start to matter and why some investors choose to pay for protection even when, on paper, it lowers expected returns.

The ``catch-up" trap: how drawdowns tempt you into bad bets

After a big loss, the portfolio doesn't just need to recover; it needs to recover *fast enough to satisfy the original plan*. That pressure can be subtle. It shows up as a portfolio drifting toward more aggressive assets, more concentrated ideas, more leverage, more "high conviction."

The investor tells themselves a story: *I'm just getting back to where I should have been.*

But the market doesn't recognize "should have been." It only recognizes risk taken and risk rewarded. When you increase risk after losses, you're not simply reclaiming the old plan—you're creating a new plan with a different probability distribution, often one that includes a non-trivial chance of making the hole deeper.

This is one of the most pragmatic arguments for drawdown control. It isn't just about avoiding pain; it's about avoiding the sequence of decisions pain tends to trigger.

If a hedge prevents you from feeling the need to "make it back" in a hurry, it can quietly save you from the most expensive mistake of all: changing strategies at the worst time.

Big losses turn good portfolios into bad portfolios through behavior

A portfolio can be well-designed on day one and poorly implemented over time. Implementation isn't just about trading mechanics; it's about human behavior under stress.

A classic pattern goes like this:

- A bear market hits.
- The investor holds through the first leg down.
- The investor capitulates near the bottom—selling because the pain feels infinite.
- The market rebounds.
- The investor waits for a "better entry," misses part of the recovery, then re-enters later.

The tragedy is that none of these decisions feel foolish in the moment. Each one feels like an attempt to regain control.

Deep drawdowns increase the likelihood of this pattern because they increase the intensity and duration of stress. Reducing drawdown isn't only a mathematical benefit; it's a behavioral intervention. Keeping a portfolio within a tolerable loss range can be the difference between "I can stick with this" and "I need to do something."

You don't need perfect discipline. You need enough discipline to avoid the handful of irreversible decisions that do the most damage.

So what are you really insuring against?

It's tempting to think of downside protection as a bet against the market. But for long-term investors, the real risk often isn't "stocks will be down this year." The real risk is a plan-breaking drawdown that causes one of these outcomes:

- you sell at the wrong time,

- you can't fund a near-term liability,
- you take on more risk than you can handle to catch up,
- you permanently lower your equity exposure after the crash and never fully re-risk.

Those are the compounding killers. Not every bear market triggers them, but big losses make them much more likely.

This is why the math of recovery belongs at the center of any serious risk conversation. It puts a price tag on drawdowns. It turns "50% down" from a dramatic headline into a concrete requirement: *100% up just to get back to zero.*

And once you internalize that, you start to see why many "reasonable" investors become obsessed with avoiding deep drawdowns. They're not trying to avoid discomfort. They're trying to avoid losing years.

Actionable takeaway: pick your maximum tolerable drawdown, then design around it

Write down a number—one number—that represents the largest portfolio decline you can tolerate *without* changing your plan, your spending, or your strategy. Don't choose the number that sounds brave. Choose the number you can live through without doing something irreversible.

Then do two simple checks:

1. Look at your current allocation and ask, *"In a severe bear market, is it plausible I will exceed this drawdown?"* If the honest answer is yes, you are not prepared for the risk you're taking.

2. Ask, *"If I exceeded it, what would I do?"* If the answer involves selling, delaying a goal, or taking bigger risks to catch up, you've found the real reason downside protection can be worth paying for.

Once that maximum drawdown is explicit, protection stops being a vague desire and becomes a design constraint. That's where smart hedging begins: not with a product, but with a limit.

1.4 The Timing Trap: Sequence of Returns Risk

Two retirees can do everything "right," earn the same long-run return, and still end up living in different financial climates—one in perpetual ease, the other in permanent caution. The difference isn't intelligence. It isn't discipline. It isn't the fund lineup. It's the calendar.

Markets don't just deliver returns; they deliver them in an order. And once your portfolio starts writing checks—monthly withdrawals, quarterly distributions, required minimum distributions, tuition bills disguised as "family help"—the order becomes destiny.

Sequence-of-returns risk is the name for this timing problem. It sounds technical. It's not. It's the simple fact that a bad early stretch can do damage that a good later stretch cannot fully repair, because you were selling shares when they were cheap. The portfolio doesn't just lose value; it loses *future earning power*.

Averages lie because you don't live in an average year

The market loves to advertise itself with averages: "stocks returned about X% over the long run." That statement is true in the way "the average depth of a river is four feet" is true—comforting right up until you step into the ten-foot trench.

Averages are silent about the path. They don't tell you whether the first few years of retirement were calm or chaotic. They don't tell you whether you were forced to sell after a 30% drawdown or whether you got to harvest gains before the storm. They don't tell you what happened to your share count—the thing that ultimately determines how much income your portfolio can produce.

When you're accumulating, the path matters less because your paycheck is still flowing in. You're adding new money, buying more shares, and downturns can even help by lowering prices during your buying years. When you're withdrawing, the path matters more because your portfolio is now the paycheck. You're subtracting shares, and subtracting shares at the wrong price is the financial equivalent of chopping down fruit trees to eat the apples.

The hidden villain is not volatility. It's *withdrawals plus volatility.*

Volatility on its own is survivable for a long-term investor who can truly wait. That's why broad equity investing works at all. But add withdrawals, and volatility becomes a different animal.

Withdrawals create what mathematicians call a "negative cash flow." In plain English: money leaves the portfolio every month whether markets are up or down. That turns a temporary decline into a permanent decision, because you must sell something to raise cash.

If prices are down when you sell, you liquidate more shares to meet the same spending need. Those shares are gone forever. They can't participate in the recovery. So even if the market rebounds strongly later, the portfolio is trying to recover with fewer oars in the water.

This is why a retiree can experience something that feels like a cruel paradox: the market "came back," but their portfolio never did.

A tale of two retirees (same returns, different lives)

Imagine two investors who retire with the same $1,000,000 portfolio and withdraw $50,000 at the end of each year (ignore taxes for simplicity). They both experience the same set of market returns over the next five years: two great years and three terrible years. The only difference is the order.

Retiree A (bad early, good later):

- Year 1: -20%
- Year 2: -15%
- Year 3: -10%
- Year 4: +25%
- Year 5: +30%

Retiree B (good early, bad later):

- Year 1: +25%
- Year 2: +30%
- Year 3: -20%
- Year 4: -15%
- Year 5: -10%

Average return? Identical. Same five numbers. Same "long-term" story.

But the lived experience is different. Retiree B spends the first two years harvesting gains. Withdrawals come from a portfolio that's swelling, not shrinking. Retiree A spends the first three years selling into declines. Their withdrawals don't just fund lifestyle-they deepen the drawdown by forcing more share sales at depressed prices.

By the time the good years arrive for Retiree A, the portfolio is fighting with a smaller base. The recovery is real, but it's trying to fill a deeper hole with a smaller shovel.

This is the timing trap: you can be right about the market's long-run generosity and still get hurt because your personal sequence was unlucky.

Why the first five years of retirement feel like walking on thin ice

Retirement is often described as a finish line. In practice, it's a handoff. You stop relying on wages and start relying on portfolio mechanics. The first few years are especially fragile because the portfolio is at its largest nominal size and the withdrawals begin immediately.

A bad early sequence does three things at once:

- It reduces the portfolio value.
- It forces withdrawals from a depressed base.
- It often coincides with a rough economy, increasing anxiety and reducing flexibility.

This is why "I can tolerate volatility" can be true at age 45 and false at age 65, even if your personality didn't change. Your financial system changed. Volatility moved from being unpleasant to being *structurally relevant*.

And it's why the earlier math of big losses becomes more than trivia here. A 50% drawdown is always painful, but in early retirement it can be catastrophic because it can require a doubling just to break even—while withdrawals are still draining the portfolio in the background. You're not climbing the Recovery Hill; you're climbing it with a backpack full of rocks.

Accumulation isn't immune: ``retirement date risk" is sequence risk in disguise

Sequence-of-returns risk isn't only a retiree problem. It can ambush investors late in accumulation—especially those close to a planned retirement date.

If you're 35 and the market drops 30%, it's emotionally unpleasant, but you have time and contributions ahead of you. If you're 62 and planning to retire at 65, the same drop can delay retirement by years. You can't "average it out" because your date isn't flexible, or because your willingness to keep working isn't infinite.

This is the version of sequence risk that shows up as a quiet life shift: *I thought I was three years away. Now I'm not sure.*

Late accumulation is also when investors often reach their maximum balance. That means the dollar impact of market swings is the largest right when your time to recover is shrinking. It's the financial equivalent of learning to surf in bigger waves as you get closer to the rocks.

The portfolio isn't just a pile of money. It's a machine with moving parts.

A portfolio that's in withdrawal mode has gears:

- a withdrawal rate (how much you take),
- a return path (the order of gains and losses),
- and a spending flexibility (whether withdrawals can drop in bad years).

Change any one of those gears, and the outcomes can change dramatically.

This is why retirees sometimes get blindsided by advice that sounded airtight during accumulation: "Just hold a diversified portfolio and don't touch it." Retirement requires touching it. The touching is the whole point. You can't pretend the mechanics are the same.

It also explains why reducing volatility can be mathematically similar to increasing return for a retiree. If lower volatility reduces the probability of deep early drawdowns, it can preserve the share base and keep the machine functioning. A smoother path can support higher sustainable withdrawals than a jagged path, even if the average return is the same.

The ``sell low" problem is not a moral failure. It's arithmetic.

People often treat selling into a downturn as a behavioral mistake. Sometimes it is. But in retirement, it can be a mechanical necessity.

If the mortgage is due or the grocery bill is real, you sell what you must sell. In a drawdown, that means selling more shares for the same dollars. That's not panic. That's arithmetic.

The portfolio doesn't care whether you feel calm when you do it. The damage is the same: fewer shares remaining to participate in recovery.

This is why the most effective retirement risk management often looks boring:

- holding enough cash to avoid selling during a downturn,
- keeping a bond sleeve that can fund spending in bad equity years (when it behaves as hoped),
- using a spending rule that reduces withdrawals after poor returns,
- or creating a baseline of guaranteed income so withdrawals from the portfolio become optional rather than mandatory.

None of this requires predicting the market. It requires admitting that the market's timing is unpredictable and building a plan that can tolerate bad timing.

A practical way to think about withdrawal rates: they're not ``safe,'' they're ``fragile'' or ``robust''

The famous question—"What's a safe withdrawal rate?"—can trick people into thinking there's a magic number. In reality, withdrawal success depends on the path of returns, inflation, fees, and personal spending behavior.

A withdrawal plan is *fragile* if it works only when the early years are kind. It is *robust* if it can survive ugly early sequences without forcing a painful, permanent lifestyle downgrade.

Robustness often comes from flexibility. A household that can reduce spending 10–15% in bad years has a materially different risk profile than a household with fixed obligations. The market doesn't negotiate with fixed obligations. It does, however, negotiate with discretionary travel, gifting, or the timing of big purchases.

That doesn't mean retirees should live in fear. It means the cleanest form of protection might be a pre-committed spending rule rather than a complex hedge.

The best time to decide how you'll cut spending is when the market is calm. The worst time is when every headline is screaming and your portfolio just fell six figures.

Why ``bucket strategies'' feel intuitive (and what they're really doing)

Many retirees like the idea of "buckets": a cash bucket for near-term spending, a bond bucket for medium-term needs, and a stock bucket for long-term growth. Critics sometimes mock this as mental accounting. But the emotional appeal exists for a reason: it maps onto the mechanical problem.

What buckets are really doing is separating time horizons inside the same household. They create a zone of money that is not exposed to immediate equity volatility so you can avoid selling stocks in a drawdown.

In other words, buckets are not about maximizing return. They are about controlling *forced selling*. They're a form of sequence-risk management, whether or not the investor uses that label.

A well-built withdrawal plan doesn't have to use buckets explicitly, but it does need to answer the same question: *Where does next year's spending come from if markets are down 30%?*

If the answer is "stocks," you are betting your lifestyle on the market's short-term mood. That can work—until it doesn't.

Hedging, re-framed: reducing the chance of a plan-breaking early sequence

When investors hear "hedging," they often imagine dramatic moves: shorting the market, buying expensive options, making bold calls. But in the context of sequence risk, hedging is often quieter. It's any deliberate design choice that reduces the probability of a plan-breaking drawdown early in the withdrawal period.

That could include a more conservative allocation near retirement, then gradually increasing equity exposure later (when the portfolio has survived the most fragile window). It could include holding a larger cash reserve than feels "optimal" in a spreadsheet. It could include partial annuitization to cover essential spending. It could include a defined-cost tail hedge sized to matter, not as a profit engine but as a shock absorber for the specific scenario that breaks your plan.

The key is specificity. Sequence risk isn't "the market might go down." It's "the market might go down *early*, while I'm withdrawing." The defense should match that.

Actionable takeaway: write a ``bad first year" plan-one page, no heroics

Before you retire (or if you're already there, before the next downturn), write a one-page plan that answers this scenario:

> *If the market drops 30% in my first year of withdrawals, exactly where will my next 12 months of spending come from, and what will I cut or pause to reduce portfolio withdrawals?*

Make the answers operational:

- which accounts get tapped first,
- what gets sold (cash, bonds, dividends, equities),
- what expenses get reduced temporarily,
- and what changes are *not* allowed (like abandoning the plan entirely or taking on leverage to "recover faster").

If you can't write that page with confidence, you don't have a withdrawal strategy yet-you have a hope strategy. Sequence-of-returns risk punishes hope, not because you were foolish, but because the calendar is indifferent. A plan that can survive a bad first year is a plan that can survive almost anything.

1.5 Insurance vs. Gambling: The Vital Distinction

The easiest way to spot confusion about hedging is to listen for tone. Insurance talk sounds like grown-ups making unromantic decisions: "We need to be covered." Gambling talk sounds like someone trying to feel clever: "This is the trade." Unfortunately, the same instrument can support both mindsets. A put option can be a seatbelt or a lottery ticket. The difference isn't the contract. It's the intent, the sizing, and the willingness to pay a cost for something you hope never happens.

Markets reward ambition often enough to make people forget that risk management is not supposed to be thrilling. A well-designed hedge is, on purpose, a little disappointing most of the time. If it's exciting in normal years, it's probably not a hedge—it's a view.

A hedge has a job description; a gamble has a storyline

A real hedge starts with a job description: *Protect the plan if X happens*. It's boringly specific. It names the scenario you can't afford-an equity crash, a credit event, a sharp rise in rates, a drawdown early in retirement-and it defines what "success" looks like.

A gamble starts with a storyline: *I think the market is about to...* It's a prediction dressed up as a plan. The "success" condition isn't survival; it's being right.

This is not a moral judgment. Plenty of investors speculate. Markets need speculators. The problem is when you *believe* you're buying insurance while you're actually buying a forecast. That's how people end up shocked when their "hedge" loses money exactly when they expected it to save them.

A homeowner doesn't buy fire insurance because they think the kitchen will ignite next Thursday. They buy it because if the house burns down, the consequences are unacceptable. The logic is consequence-first, not probability-first. Hedging should feel the same.

The three tells: why most "hedges" aren't hedges

Many strategies get called hedges because the word feels responsible. But you can usually tell what you're looking at by asking three simple questions.

Tell #1: Does it have a defined cost? Insurance has a premium. A hedge should have a cost you can name, budget, and live with.

If a strategy is sold as "free protection" or "no-cost hedging," you're being invited into a trade-off you may not understand yet. Sometimes the cost is hidden: capped upside, exposure to gaps, left-tail risk, leverage, or path dependency. If you can't explain the cost in plain English, you don't own insurance-you own complexity.

Tell #2: Does it target a specific risk? A hedge is directional in the sense that it responds to a particular bad outcome, but it is not a broad bet that "the market will fall."

For a long-term investor, "the market might fall" is not a thesis. It's a weather report. The hedge needs to be tied to something that matters to your life: a drawdown beyond a threshold, a retirement-start crash, a concentrated position blow-up, a funding ratio breach, a margin call risk, a liability coming due.

A strategy that "might make money in a downturn" is not automatically a hedge. A hedge has to show up *when you need it*-not merely in some down markets, but in the kind of down market that breaks your plan.

Tell #3: Is it sized to matter, or sized to entertain? A hedge that's too small is like a smoke detector with dead batteries. It provides psychological comfort until the moment it's asked to do real work.

Sizing is where intentions reveal themselves. Speculative trades are often sized for excitement ("If I'm right, this pays!"). Insurance is sized for continuity ("If I'm wrong about timing, I can still live with the cost; if the disaster hits, I'm covered enough to keep the plan intact.").

If the hedge is so small that it can't meaningfully reduce drawdown, prevent forced selling, or support your spending plan, it's not protection. It's a talisman.

Options: the instrument most often mistaken for a personality

Options are the poster child for this confusion because they can be used in so many ways. The same contract type can create:

- a defined-cost floor under a portfolio (insurance),
- extra income in calm markets (which is often selling insurance),

- leverage for directional bets (speculation),
- or complicated payoff diagrams that look impressive in a backtest (performance art).

Options themselves are neither good nor evil. They are tools with sharp edges. Exchange-traded options also come with formal disclosures for a reason: they can behave in unintuitive ways, and they are not suitable for every investor or every situation.

A protective put-buying the right to sell at a fixed price-is the cleanest insurance analogy in markets. You pay a premium. If the market falls below the strike, the put can gain value and soften the blow. If the market rises, the put expires and you "wasted" the premium in the same way you "waste" homeowner's insurance in a year when your house doesn't burn down.

The problem is that many investors buy puts the way people buy lottery tickets: infrequently, emotionally, and with a vague hope of a dramatic payout. That's not hedging; that's buying catastrophe exposure and calling it prudence.

Insurance works best when it's deliberate and repeatable, not reactive and theatrical.

The premium is the point: why good insurance feels like a drag

The market is not a charity. If there is a reliable way to profit from fear, someone will monetize it. In practice, investors who want crash protection often pay a premium for it-sometimes explicitly (option premiums), sometimes implicitly (holding lower-return assets, giving up upside, paying for structured products).

This is why many hedges have negative carry: they tend to lose small amounts in calm markets. That loss is not a "failure." It's the premium.

People who can't tolerate paying the premium are often tempted to reshape the hedge until it no longer costs anything. That's when they accidentally become the insurer.

Selling options for income, for example, can feel like getting paid for your patience. And sometimes it is a sensible strategy within limits. But it's also economically similar to collecting insurance premiums-steady nickels most of the time, with the risk of handing back dollars in a crisis. The mirror-image of buying insurance is selling it. Neither is inherently wrong. Confusion is what's dangerous.

If you want the hedge to behave like insurance, you must accept that it will behave like insurance: a recurring cost that occasionally becomes priceless.

``I hedged and still lost money"---a statement that can be either wise or foolish

People often judge hedges the way they judge trades: did it make money? That's the wrong scoreboard.

The better question is: *Did it do its job?*

If the market rose 20% and your hedge lost 2%, you did not "lose" 2%. You paid 2% for protection you didn't need. That can be a perfectly rational outcome if the 2% was budgeted and if the hedge would have meaningfully reduced damage in a crash.

If the market fell 35% and your hedge was down too-or barely helped-then you didn't just have a bad year. You had a design problem. Either the hedge didn't match the risk, or it was sized incorrectly, or it had hidden path dependencies, or it relied on correlations that didn't show up.

This is where many investors learn an uncomfortable lesson: a hedge can be "reasonable" and still fail *for you* because the scenario you needed protection from wasn't the scenario it was built to address.

Buying inflation protection doesn't help in a deflationary crash. Buying a short-term put doesn't help if you keep rolling it at the wrong times. Owning low-volatility stocks doesn't protect you from equity-wide deleveraging. Calling something a hedge doesn't make it one.

The most common form of gambling disguised as hedging: market timing with a safety label

There is a particularly seductive habit among investors: taking a bearish view, then labeling it "risk management" so it sounds disciplined.

It usually looks like one of these:

- Going to cash because "things feel frothy," with no clear re-entry rule.
- Shorting the market because "a recession is coming," without defining what would prove the thesis wrong.

- Buying far-out-of-the-money puts because they're cheap, hoping for a crash-sized payoff.
- Rotating aggressively among assets based on macro predictions, calling it "dynamic hedging."

Sometimes these moves work. That's what makes them addictive. But their success depends on being right about timing—*and* often being right twice: when to get defensive and when to get back in.

Insurance doesn't require being right twice. It requires being solvent when the bad thing happens.

A hedger can be wrong about timing for years and still be "correct" if the cost is controlled and the hedge is there when the disaster arrives. A market timer who is early is not "early." They are wrong until the moment they are right, and the cost of being wrong can be enormous.

This difference—needing to be right once versus needing to be right twice—is why hedging and gambling feel similar in a brokerage account but behave very differently in a life.

A hedge should reduce regret, not maximize bragging rights

Speculative trades are social. They come with stories. You can tell friends you "called it." Hedges are lonely. If they work, you feel relief, not pride. If they don't pay out, you feel mildly annoyed but also strangely grateful.

That emotional signature is a clue.

The purpose of protection is to reduce the chance of a plan-breaking decision: panic selling, forced selling, abandoning retirement timing, cutting support for family, returning to work under duress. A hedge is successful when it makes your future self's life less brittle, not when it generates cocktail-party returns.

Ironically, the best hedges can look stupid right up until the day they become the only thing in the portfolio that feels sane.

Design principles that keep hedging on the "insurance" side of the line

If you want to stay out of the gambling zone, a few design principles do most of the policing:

Define the insured event in one sentence. Not "the market might go down," but "I want to limit my portfolio drawdown to roughly X% over Y months so I can continue withdrawals without selling equities at depressed prices."

Budget the premium like a bill. If you can't tolerate the ongoing cost in good markets, you will abandon the hedge before it matters. Insurance you cancel is just expensive stress.

Measure success by drawdown reduction, not profits. If the hedge reduces a 40% decline to 25% in the scenario that would otherwise break your plan, it did its job-even if it didn't "make money" in a rising market beforehand.

Avoid strategies that require perfect timing. If the hedge must be initiated or removed at exactly the right moment, it's not insurance; it's a forecast wearing a trench coat.

These principles aren't glamorous, which is exactly why they work. They turn hedging from a thrilling idea into a repeatable practice.

Actionable takeaway: the "hedge litmus test" you can run in five minutes

Before you place any trade or adopt any strategy you're calling a hedge, answer these questions in writing:

1. *What specific scenario am I insuring against?* (A 30% equity drawdown? A crash in the first two years of retirement? A concentrated position collapse?)
2. *What is the maximum annual cost I am willing to pay for this insurance?* (In dollars or percent-something you can actually live with.)
3. *If the market goes up for two years and my hedge loses money the entire time, will I keep it?* (If the honest answer is no, you don't have insurance; you have a bet.)

If you can answer those cleanly, you're thinking like an insurer: consequence-first, cost-aware, and humble about timing. If you can't, pause. The market will happily let you speculate. It won't always let you pretend it was risk management.

1.6 The Price of Sleep: Understanding Opportunity Cost

The cheapest form of protection is the one nobody wants to buy: doing less. Less leverage. Less concentration. Less exposure to the hottest corner of the market. Less bravado. In a bull market, restraint looks like a personality flaw. In a bear market, it looks like foresight.

Hedging, in any form, asks you to accept a trade: you give up some upside (or pay a recurring premium) to reduce the probability of a plan-breaking hit. That trade is the *price of sleep*. Some investors pay it gladly. Others refuse on principle-until the night the smoke alarm goes off.

Opportunity cost is the language of that trade. It's what you *don't* earn because you chose to defend. And it's easy to misunderstand, because the cost is usually visible and immediate while the benefit is invisible and conditional.

Opportunity cost is not a fee; it's a life choice in disguise

A management fee shows up on a statement. Opportunity cost does not. That makes it psychologically slippery.

If you buy a protective put and it expires worthless, you can point to the premium and feel it as a "loss". If you hold extra cash and the market rallies, you can look at the gap versus a fully invested portfolio and feel behind. If you tilt toward bonds and equities surge, you can feel like you made a mistake.

But these feelings confuse two different games:

- the game of maximizing returns in good markets, and
- the game of avoiding ruin in bad markets.

Most people are playing both games at once, whether they admit it or not. The problem is that the scoreboards are different. In the first game, opportunity cost is a glaring number. In the second game, the "profit" is that your life didn't break—and that profit rarely prints on a statement.

A hedge is not meant to outperform in sunshine. It is meant to keep you from being crushed in a storm. Complaining that an umbrella didn't help you tan is missing the point.

The most honest sentence in hedging: ``This will probably cost me money."

The market charges for certainty. If protection were consistently cheap, it would be arbitraged away. If it were consistently profitable, everyone would do it until it wasn't.

That's why many forms of downside protection have negative carry-the expected experience of losing small amounts over time in exchange for occasional large benefits. Think of it as paying rent on stability.

This is especially clear with options-based hedges. If you buy insurance against a crash, and crashes are rare, you will usually lose the premium. That is not a bug; it's the business model. The seller collects premiums precisely because most periods are normal.

Even outside of options, the same logic applies. Holding extra cash can reduce your long-run expected return because cash typically yields less than equities over long periods. Holding more bonds can reduce expected return. Owning "defensive" assets can mean paying up for safety, which can reduce future returns if you overpay.

The cost is real. The question is not whether you pay. The question is whether you're paying for the right thing.

The wrong way to think about cost: ``Did the hedge make money?"

A common mental trap is to judge hedges the way you judge investments: by whether they produced positive returns.

That framing will turn you into an inconsistent hedger-buying protection after volatility rises (when it's expensive), abandoning it after markets calm (when it's cheap), and repeating the cycle like a seasonal allergy sufferer who takes medicine only after the pollen counts spike.

A hedge is better evaluated the way you evaluate a fire extinguisher:

- Did you have it when you needed it?
- Was it big enough to matter?
- Did it work as expected under stress?

The "return" is not measured in dollars earned during calm years; it's measured in damage avoided during the years you can't afford to repeat.

If that sounds too abstract, translate it into the only currency that ultimately matters: *choices*. A hedge is valuable if it preserves choices-keeping you from being forced to sell, forced to cut spending drastically, forced to delay retirement, forced to concentrate risk to catch up.

The most expensive cost is the one you don't see: forced behavior

Opportunity cost often gets reduced to a number: "My hedge cost 2% this year." But the truly expensive cost is often behavioral and only becomes visible in hindsight.

A deep drawdown can force you into opportunity cost of the worst kind:

- Selling at low prices, then missing the recovery.
- Holding too much cash for years afterward because you no longer trust the market.
- Taking on higher risk later to "make it back," increasing the chance of a second disaster.

These are not minor side effects. They can dwarf the explicit premium of almost any sensible hedge.

This is why the "price of sleep" should be compared not to an idealized fully invested portfolio in a perfect world, but to the cost of a bad decision in a real one. If a hedge costs you a predictable 1–3% per year but prevents a once-in-a-decade panic exit that costs you 20% in missed recovery, the hedge wasn't "expensive". It was a bargain.

Of course, hedges don't always work that cleanly. But the comparison set matters. If you benchmark insurance against perfection, it will always look wasteful. If you benchmark it against human behavior under stress, it often looks rational.

Explicit versus implicit costs: you're paying either way

Investors sometimes avoid hedging because they dislike paying visible premiums. That aversion is understandable-and often misplaced, because many "unhedged" portfolios pay implicit premiums anyway.

Explicit cost:

- Option premiums for protective puts or collars.
- Fees embedded in structured products or hedge funds.
- Transaction costs and slippage from frequent rebalancing or rolling positions.

Implicit cost:

- Holding cash instead of equities (foregone expected return).
- Holding more bonds than you otherwise would (foregone expected return, and sometimes inflation risk).
- Owning defensive assets at rich valuations (paying up for perceived safety).
- Concentrating in "safe" income strategies that are actually short-volatility in disguise (collecting pennies, risking dollars).

The market is happy to charge you in whatever form you prefer: a premium you can see, or a return you never earned.

There is no free lunch-only different receipts.

Inflation: the quiet opportunity cost hiding inside ``safety"

Safety is not a single dimension. Many investors treat "not losing money" as the definition of safe. But money that sits still can still lose purchasing power.

Inflation is the slow leak in the "safe" bucket. It's not dramatic enough to trigger panic, which is precisely why it's dangerous. If you hold large amounts of cash or very short-term instruments for long periods, you may be reducing volatility while quietly increasing the risk that your money buys less of your future life.

This matters because many hedging decisions are really time-horizon decisions. A retiree with near-term spending needs may be right to hold more cash-cash buys flexibility and prevents forced selling. But a 40-year-old holding a large "safety" allocation for emotional comfort may be paying an inflation tax for decades.

The right amount of safety depends on what the money is *for* and when you need it. Safety is not "low volatility." Safety is "high probability of meeting the goal."

The volatility risk premium: why insurance is usually priced to disappoint buyers

Markets have a well-known habit: people are willing to pay up for protection. They dislike large losses more than they enjoy equally sized gains. That human preference shows up in option prices.

In practice, this can mean that the implied volatility baked into options tends to be higher than the volatility that actually materializes. Said plainly: crash protection is often priced like it's more likely than it turns out to be. The buyer pays for fear. The seller collects for courage.

This doesn't mean buying protection is foolish. It means you should enter with eyes open. A recurring hedge can be a purposeful drag-like paying for a security system you hope never proves "worth it". If you treat it like a profit center, you will be disappointed. If you treat it like catastrophe planning, you may be grateful.

The market doesn't offer emotional comfort at wholesale prices.

The ``sleep well" budget: decide what you can pay before you shop

One reason investors overpay for hedging is that they shop for it in the wrong order. They start with an instrument ("puts," "tail-risk fund," "buffer ETF," "collar strategy") and only later ask, "How much will this cost me?"

It's more rational to reverse the sequence:

1. Decide the maximum annual drag you can tolerate.
2. Decide the specific scenario you're insuring against.
3. Then choose a tool that fits both.

This is how every other kind of insurance works. You don't walk into an insurance office and say, "Sell me the most sophisticated policy you have." You decide what you need covered, you decide what you can pay, and you accept that beyond a certain point, more coverage costs more.

The same discipline protects you from the two classic hedging errors:

- *Over-insuring*: paying so much premium that you permanently reduce your odds of long-term success.
- *Under-insuring*: buying something so small or so misaligned that it won't matter when the bad event arrives.

The sweet spot is rarely "maximum protection." It's "enough protection to keep the plan intact."

When opportunity cost is actually a feature

Here's a counterintuitive but useful thought: opportunity cost can be the *proof* that your hedge is doing its job.

If you buy protection and markets grind higher in a calm, low-volatility regime, your hedge will likely look like a small, steady loser. That's the expected outcome. That's why it was available to buy. That's why it can pay off when the regime changes.

The investor who can calmly pay that cost is the investor who is most likely to have protection in place when it matters. The investor who demands that the hedge "pay for itself" is often demanding an impossibility-or demanding a strategy that quietly introduces other risks.

Wanting insurance to be profitable is like wanting seatbelts to be comfortable enough that you forget they exist. The belt should remind you, gently, that you made a deliberate trade: a little discomfort for a lot of survivability.

A practical way to compare costs: ``drag versus disaster"

Opportunity cost becomes easier to evaluate when you put it into a simple comparison.

Suppose a protection approach costs 2% per year in expected drag. That sounds painful-until you remember what it's trying to prevent.

If the protected scenario is a 40% drawdown that would force you to sell, or delay retirement, or slash spending, then 2% per year might be cheap. If the protected scenario is a mild correction that you can live through easily, then 2% is probably too much.

The correct question is not "Is 2% too expensive?" The correct question is "Too expensive *relative to what*? What is the dollar value of avoiding the disaster scenario for my specific life?"

That is a personal calculation, not a market statistic.

And it's why two investors can rationally make opposite choices. One investor may have a pension, low spending needs, and decades of runway; hedging may be an unnecessary drag. Another investor may have tight margins, early-retirement withdrawals, and a plan that breaks at a certain drawdown; hedging may be the most rational expenditure in the budget.

Actionable takeaway: set your annual ``protection premium" like you set a savings rate

Pick a number you are willing to spend-every year-on downside protection, expressed as a percentage of the portfolio (for example, 0.5%, 1%, or 2%). This is your *sleep well budget*. Then write two lines underneath it:

1. *The scenario I'm buying protection for is:* (define the drawdown or timing risk that breaks your plan).
2. *If this premium feels painful in good markets, I will remind myself that:* (name the exact bad decision the hedge is meant to prevent-panic selling, forced selling, delaying retirement, or cutting essential spending).

If you can't name the scenario and the prevented decision, don't pay the premium. If you can, and the premium is within your budget, you've turned hedging from a vague fear response into a deliberate household expense-like insurance should be.

Chapter 2

The Hedger's Toolkit: Instruments and Mechanics

Financial derivatives often have a reputation as 'Weapons of Mass Destruction,' but for the prudent investor, they are simply precision power tools. Just as you wouldn't try to frame a house with a rock, you can't build a robust hedge using only stocks and bonds. This chapter opens the toolbox, moving past the jargon to explain exactly how contracts like options and futures work, how to visualize their 'hockey stick' payoff curves, and which retail products (like inverse ETFs) are safe to handle versus which ones might explode in your hand if held too long.

2.1 Introduction to Derivatives: Contracts, Not Assets

Imagine your neighbor has a gorgeous old oak tree that leans a little too far over your roof. Most years it's just shade and falling leaves. But every time a storm warning pops up on your phone, you feel that little clench in your stomach: *If that thing comes down, it's coming down on me.* You could offer to buy the tree (good luck), or you could cut it down (not your property, and also: awkward). What you really want is simpler: a way to make the financial pain of "tree meets roof" someone else's problem.

That instinct—*I can't control the risk, but I can control who eats it*—is the emotional engine behind derivatives.

A derivative is not a thing you own like a share of stock or a bond certificate. It's a contract: a legal agreement whose value *derives* from something else—an underlying reference like a stock, an index, an interest rate, a currency exchange rate, a commodity price, even the weather in some specialized markets. If the underlying moves, the derivative's value moves in a rule-based way the contract spells out ahead of time. You're not buying the house; you're buying the insurance policy. You're not buying the corn; you're buying the agreement tied to the price of corn.

That distinction sounds academic until you feel it in your bones during a drawdown. Owning an asset means you live inside its day-to-day mood swings. A contract lets you *reshape* how those mood swings hit your portfolio—sometimes by softening the blow, sometimes by trading away upside, sometimes by converting a wild ride into a calmer one with a known cost.

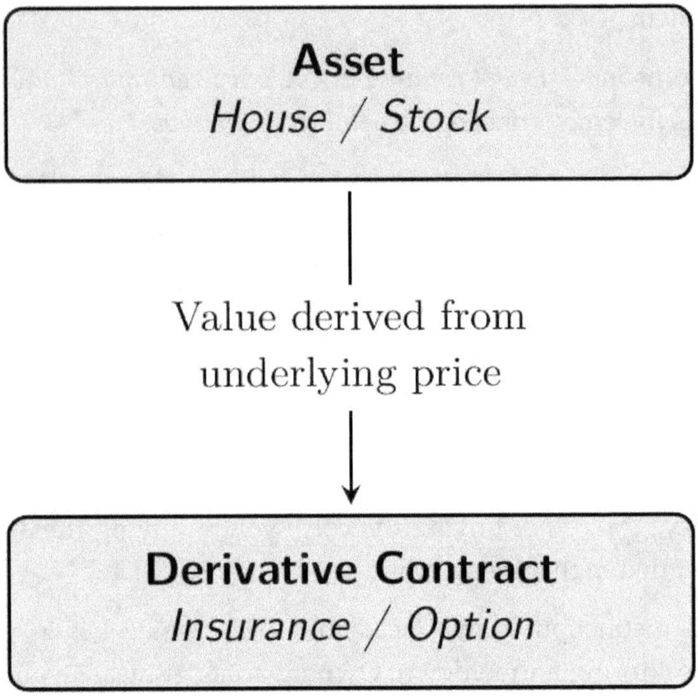

There's a subtle comfort in the word "contract." Contracts have terms. Terms have boundaries. Boundaries are what a panicking investor is usually missing. When markets are falling and headlines sound like sirens, your brain wants certainty: *How bad can this get? What do I do if it gets worse?* A derivative can answer those questions in advance—not perfectly, not for free, but explicitly. The market can still be chaotic; your payoff doesn't have to be.

Derivatives exist because different people experience the same price move as different kinds of pain. A wheat farmer hates the price falling between planting and harvest. A cereal company hates the price rising. A pension plan hates interest rates falling because future obligations become more expensive. A homeowner hates hail. In each case, someone is exposed to a risk they didn't sign up for emotionally, even if they signed up for it economically. Derivatives are the plumbing that lets one party pass a specific risk to another party willing to hold it—often because that other party has an offsetting exposure, a different time horizon, or simply a higher tolerance for the roller coaster.

If that sounds like "insurance," good. Insurance is one of the cleanest everyday derivatives we all understand. Your auto policy's value depends on something you hope never happens. You don't "own" a car accident. You own a contract that changes what a car accident does to your finances. In investing, derivatives do the same thing with market accidents.

The "side-bet" framing is useful because it strips away mystique. A derivative is a side-bet on the behavior of an underlying reference. Your brokerage screen might show it next to your stocks, but it's a different species. A stock is a slice of a business. A bond is a claim on cash flows. A derivative is a set of *if-then* statements with a price tag: if the S&P 500 is below X at time Y, then pay me Z (or let me sell at X, or make up the difference). That's why derivatives can be both humble and dangerous. Humble because they're just rules. Dangerous because rules can create leverage, and leverage can turn a modest move into a life-changing loss if you're on the wrong side of the contract.

The best way to keep your footing is to ask a derivative one blunt question: *Does this contract give me a right, or does it give me an obligation?* Rights are what you buy when you want protection. Obligations are what you sell when you want income or cheaper protection—while taking on the duty to perform if the market moves against you. The market is full of people eager to sell obligations in quiet times because it *feels* like getting paid for nothing. The market is also full of people who discover, too late, that "nothing happened today" is not a risk management plan.

Two big families of derivatives cover most of what long-term investors encounter.

The first family is the forward-commitment world: forwards, futures, and swaps. These are agreements to transact later, and they behave like straight-line exposure. If the underlying moves against you, the contract moves against you, typically dollar-for-dollar in a way that can feel brutally honest. These tools are powerful for adjusting your portfolio's overall exposure—turning the dial down or up—but they don't magically create a safety net. They're more like replacing your car with a different car: still a car, still on the same road, just with different horsepower.

The second family is the contingent-claim world: options. Options are closer to insurance because they can create asymmetry. They can be structured so that you pay a known cost up front and, in return, you get protection only if the bad thing happens. This is why investors gravitate toward options for "crash protection": the whole point is to suffer less in the ugly tail of the distribution, even if that means giving up a little performance in normal times.

Even before you learn any vocabulary—strike prices, expirations, premiums—one mental model does a lot of heavy lifting: *derivatives are about customizing the shape of your outcomes*. You already customize your portfolio by choosing stocks versus bonds, U.S. versus international, growth versus value. Derivatives let you customize beyond that: you can redraw where your pain begins, how steep it gets, and whether you have a hard floor or a soft one. They are outcome design tools. The trade-off is that outcome design requires precision, and precision demands you respect the mechanics.

Respecting mechanics starts with accepting a mildly uncomfortable truth: a derivative is not automatically "safer" because it's not the underlying. Many derivatives are leveraged by nature. Not because someone is trying to trick you, but because contracts let you control a large *notional* exposure with a relatively small amount of cash changing hands today. This can be wonderful when used deliberately—like using a seatbelt instead of building a tank. It can also be catastrophic when used casually—like driving a sports car the way you drive a sedan.

Here's the intuition. If you buy $10,000 of an index fund, you have $10,000 of exposure. If you enter a derivative tied to $100,000 of that index, you can get $100,000 of exposure with far less than $100,000 invested. That's leverage. Sometimes leverage is explicit (margin). Sometimes it's embedded (the contract's sensitivity to price moves). Either way, the contract can move your P&L faster than the underlying would if you simply owned it. That speed is not "advanced." It's just physics.

The other mechanical reality is that derivatives are priced, not gifted. There is no free hedge. If a contract is likely to pay you in disasters, someone is charging you for that privilege—either as an upfront premium, or as worse pricing, or as the opportunity cost of giving up some upside. Investors often hate this because paying for protection feels like "wasting money" right up until the moment it is the only line item in the portfolio that is green.

A clean way to think about the cost is to compare it to home insurance again. If your house never burns down, the insurance premiums can look like a decades-long bad deal—until you remember what you were buying: not profit, but survivability. Hedges are rarely performance enhancers. They are regret reducers. They're there to keep a bad market from forcing you into a bad decision.

Where derivatives trade also changes the feel of the risk. Some derivatives are exchange-traded: standardized contracts on regulated exchanges, where a clearinghouse sits in the middle, collects margin, and tries to make sure the system doesn't collapse if one participant can't pay. Standardization can be boring, and boring is underrated in risk management. Other derivatives are over-the-counter (OTC): privately negotiated agreements between two parties, customizable and sometimes elegant, but with more dependence on the other party's ability to perform. Customization is seductive. It can also hide complexity in the fine print, where risk loves to live.

If you're a long-term investor, you don't need to become a market-structure expert to benefit from derivatives. But you do need to cultivate a "contract reader" mindset. Every derivative position can be reduced to a handful of questions that protect you from self-inflicted surprises:

What is the underlying reference? (A stock? An index? A volatility measure?)
What triggers a payoff? (A move below a level? Above a level? A certain date?)
Do I have a right or an obligation?
What's the maximum loss? (If the answer is "uncapped," your position size had better be tiny and intentional.)
What's the maximum gain? (If you're giving away the upside, is that a trade you can live with?)
What can force me to add cash at the worst time? (Margin calls don't care about your long-term plan.)

That last question is where the human side of derivatives gets real. The most painful investing losses are not just financial; they're behavioral. It's one thing to watch your portfolio decline. It's another thing to be forced to sell into the decline because a contract demands cash immediately. Derivatives can be designed to reduce panic selling—or they can be designed (or misused) in a way that makes panic selling inevitable. The difference is usually not intelligence. It's whether the investor thought about cash flows and obligations *before* the stress hit.

A practical example makes this concrete. Suppose you own a broad stock index fund for the long haul. You're not trying to trade it. You just want to avoid the kind of drawdown that makes you abandon the plan. One approach is to sell some of the fund and hold more cash or bonds. That's not wrong; it's the simplest hedge in the world. But it changes the entire portfolio all the time, even when markets are calm.

A derivative hedge aims for something more surgical: keep the long-term position, but add a side contract that behaves differently in the states of the world you fear. In calm markets, the hedge may slowly cost you money (like an insurance premium). In ugly markets, it may pay off, offsetting part of the decline. The underlying asset remains the engine. The derivative is the roll cage.

The mistake is to treat derivatives as "extra return gadgets." They are *risk transfer devices*. Used well, they help you stay invested in the assets you actually want to own by making the ride emotionally and financially survivable. Used badly, they become the asset you didn't realize you owned: a concentrated bet with sharp edges.

One more perspective helps keep the moral compass straight: in a world without derivatives, risk doesn't disappear. It just stays stuck to whoever happens to be holding it. Derivatives are one of the few ways financial markets let people unbundle risk from ownership. That unbundling is why a farmer can focus on farming, an airline can focus on selling tickets instead of gambling on fuel, and a retiree can focus on spending rather than watching every market squiggle like it's a heart monitor.

Actionable takeaway: Before you touch any derivative, write a one-sentence "contract translation" in plain English that answers (1) what it's tied to, (2) what must happen for it to help you, (3) whether you hold a right or an obligation, and (4) your worst-case loss in dollars. If you can't do that on a blank sheet of paper, you don't own a hedge—you own a mystery.

2.2 Understanding Options Mechanics

If stocks are the house, options are the paperwork that changes what the house can do to you. The paperwork can be boring—until it saves you from a financial faceplant. Options are the most common "shape-shifting" contracts regular investors bump into, and the secret to understanding them is to stop treating them like a casino chip and start treating them like two familiar objects: an insurance policy and a coupon.

An option is a contract that gives someone a *right* for a limited time. The other side of the contract takes on the matching *obligation.* That one sentence is the whole story, and it's why options feel so different depending on whether you're buying or selling them. Buy an option and you're paying to be protected (or to lock in a deal). Sell an option and you're getting paid to *be* the protection (or to offer the deal). The market isn't "giving" you premium; it's paying you to hold a responsibility.

The two basic species are calls and puts. Everything else-spreads, collars, fancy multi-leg strategies-are just plumbing built from those two pipes.

A *call* is a coupon for a future purchase. It gives the buyer the right to *buy* the underlying at a pre-agreed price. Picture a coupon that says: "Any time until June 30, you may buy this item for $100." If the store price jumps to $140, your coupon becomes precious. If the store price stays below $100, your coupon expires in your junk drawer, a small tragedy in paper form.

A *put* is an insurance policy with a deductible-like threshold. It gives the buyer the right to *sell* the underlying at a pre-agreed price. Picture a policy that says: "Any time until June 30, if the resale value collapses, you can still sell to us at $100." If the market price drops to $60, you don't have to beg for bids-you have a contract that will take it at $100. If the market never drops, you're out the premium, just like any insurance you're thrilled not to use.

That "right" language is not poetic; it's legal. If you *buy* a call, you have the right to buy. If you *sell* (write) that call, you have the obligation to sell if the buyer uses their right. If you *buy* a put, you have the right to sell. If you *sell* that put, you have the obligation to buy if the buyer uses their right. Options are a beautifully symmetrical deal: one side holds a steering wheel, the other side holds a seatbelt—and they trade places depending on which contract you're in.

The next term that trips people up is the *strike price*. The strike is just the contract's "locked-in" price—the $100 printed on the coupon or the $100 printed on the insurance policy. When you hear "buy the 400 strike put," translate it into human: "Buy a contract that lets me sell at 400."

What does the strike *do*? It defines where the option starts to matter. A put with a strike far below today's market is like a disaster-only policy: cheap, but it only kicks in after a big drop. A put with a strike close to today's market is like low-deductible insurance: more expensive, more sensitive, and more comforting on bad days. Calls are the mirror image: a call with a strike above today's market is a coupon for a deal you hope becomes valuable; a call with a strike below today's market is like a coupon that's already "in the money," which makes it pricier because it already represents a deal.

Now bring in the most underappreciated word in options: *expiration*. Every option comes with a clock. At expiration, the contract stops being a possibility and becomes a settled reality: it either has value or it doesn't. That deadline is why options don't behave like stocks. A stock can be held through your indecision for decades. An option does not care about your feelings. It will die on schedule.

The clock changes how you experience risk. With a stock, time is your friend—at least in the sense that you can wait for mean reversion, earnings growth, and the slow grind of compounding. With a long option (an option you bought), time is a cost. The option is a melting ice cube: every day that passes is one fewer day for the market to move in the direction you need. That's why an option can be "right" about the long-term story and still lose money: you can be correct and early, and in options those are not the same thing.

The price you pay for the option is the *premium*. Think of premium as the sticker price of the contract, like an insurance premium or the cost of buying the coupon. People sometimes talk about premium as if it's a fee that disappears into the ether. It doesn't disappear; it's transferred to the seller. The buyer pays to obtain the right. The seller gets paid to carry the obligation. Options are not a machine that prints returns; they are a marketplace for transferring a specific kind of risk.

If you want to feel the logic in your hands, use a concrete mental example that resembles real investing behavior. Say a stock is trading at $100.

Buying a put with a $95 strike that expires in three months is like saying: "I'm willing to pay a known amount today so that, if this thing goes off a cliff in the next three months, I have a guaranteed exit near $95." You're still exposed to the first $5 of loss (plus the premium), but you're no longer exposed to unlimited downside during that period. It's not free protection, but it is *bounded* pain, which is sometimes the only kind of pain you can plan around.

Buying a call with a $105 strike expiring in three months is like saying: "I'm willing to pay a known amount today for the chance to participate if this stock rips higher quickly." If nothing happens, your loss is limited to the premium. That limited loss is why calls are tempting to speculators. It's also why calls can be a disciplined tool for investors who want defined risk, provided they're honest about the premium being a recurring expense if they keep trying.

Flip the position and you feel the emotional swap immediately. *Selling a put* is agreeing to buy the stock at the strike price if the buyer wants to sell it to you. Some investors use this like a "limit order that pays you": "I wouldn't mind owning it at $95, so I'll get paid a premium for taking that obligation." That can be reasonable—until the market gaps down hard and you realize you didn't just promise to buy at $95; you promised to buy *because someone else is desperate to sell*. The premium is not a reward for being clever; it's compensation for standing in front of the falling knife.

Selling a call is agreeing to sell the stock at the strike price if the buyer exercises. This is the covered call world: "I own the stock, and I'm willing to cap my upside in exchange for premium." That trade can be perfectly rational. It can also generate a particular kind of regret: watching the stock rocket past your strike and realizing you sold your future joy for a small, immediate check. That regret isn't a reason not to do it; it's a reason to price it correctly in your mind. You're not "earning income." You're selling away one kind of future.

Options also introduce the deceptively simple idea of *moneyness*, a word that sounds like it should be illegal. Moneyness just means whether the option has value *right now* if it could be used immediately.

A call is *in the money* when the stock price is above the strike (your coupon is already a deal). A call is *out of the money* when the stock price is below the strike (your coupon isn't useful yet). A put is in the money when the stock price is below the strike (your insurance would pay if used), and out of the money when the stock price is above the strike (no claim to make). There's also *at the money*, roughly when stock and strike are close-where the option is most sensitive to small moves, like a seesaw balanced on its pivot.

That sensitivity leads to the next piece of vocabulary that matters more than it sounds like it should: the difference between an option's *intrinsic value* and its *time value*. Intrinsic value is the "use it right now" value. If a stock is at $90 and you own a $100 put, that put has $10 of intrinsic value because it lets you sell for $100 something the market values at $90. Time value is everything on top of that-what you pay because there's still time left for the situation to change.

Time value is why options can feel like they have a personality. With plenty of time left, an option can be expensive even if it's out of the money, because the market is paying for possibility. As expiration approaches, possibility gets squeezed. The market stops paying for dreams and starts paying for math.

This is also where the word *volatility* sneaks into the room. Volatility, in plain English, is how violently the underlying tends to swing around. Options tend to be more expensive when volatility is higher because big swings make it more likely the option becomes valuable before expiration. If the world feels calm, insurance is cheaper. If the world feels like it's holding a lit match near gasoline, insurance gets pricey. That's not market moodiness; it's the contract being priced by the probability of needing it.

You don't need to become a volatility trader to use options for hedging, but you do need to internalize one uncomfortable thing: you often want to buy protection precisely when it is most expensive. After a scary headline, options premiums can jump. That doesn't mean the hedge is "bad." It means you're walking into the store after everyone else noticed the storm clouds. The way around that isn't genius; it's planning. You decide what you'll pay for protection *before* you're afraid.

Options trading also comes with a practical reality that people learn only after they've opened a brokerage account and clicked the wrong button: brokers and regulators treat options as complex products for a reason. There are disclosures and approval levels because the same contract can be low-risk in one form (buying a put with cash paid upfront) and high-risk in another (selling uncovered calls, selling puts without the ability to fund an assignment). The complexity isn't just the vocabulary; it's the embedded obligations.

Here's a compact "translation dictionary" that keeps you from getting hypnotized by ticker symbols and jargon:

Call	A coupon: the right to buy at the strike before expiration.
Put	Insurance: the right to sell at the strike before expiration.
Strike	The locked-in price printed on the contract.
Expiration	The deadline. After this, the contract is over.
Premium	The price paid for the contract (paid by buyer, received by seller).
Buyer / Holder	Owns a right; risk is typically limited to the premium paid.
Seller / Writer	Owes an obligation; risk can be large if the move is large.

Options can be used recklessly-people do it every day-but the mechanics themselves are not reckless. They're precise. They tell you, in advance, what you're allowed to do, what you must do, and when the whole arrangement ends. For a long-term investor trying to hedge, that precision is the point. You're not trying to predict the next headline. You're trying to decide, calmly, what you want your worst month to look like.

A good final gut-check is to say your position out loud in plain language as if you're explaining it to someone who doesn't care about finance. If it sounds like a sensible sentence, you're probably in the realm of responsible mechanics. If it sounds like a riddle, you're probably underwriting risk you don't fully see yet.

Actionable takeaway: Before placing any options trade, write two lines on paper: "My right is: __________" or "My obligation is: __________", and "This contract expires on: __________." If you can't fill those in without looking back at the trade ticket, pause. Most expensive options mistakes begin as small moments of confusion.

2.3 Visualizing Protection: Payoff Diagrams

Most investing arguments are really arguments about feelings. Someone says, "I can handle risk," and what they mean is, "I can handle risk as long as it doesn't show up all at once, in public, while I'm trying to sleep." Payoff diagrams are the antidote to that kind of vagueness. They force the conversation out of slogans and into geometry. You can look at two lines on a chart and answer, with uncomfortable clarity: *Where do I start bleeding? How bad can it get? What did I buy with the premium?*

A payoff diagram is not a price chart. It doesn't tell you what the market *will* do. It tells you what your position *does* if the market ends at different prices on a specific date—typically an option's expiration. That's a big psychological shift. Price charts are narrative bait: they tempt you to explain the past and predict the future. Payoff charts are contract truth: they show the deal you signed, whether or not the market cooperates.

The "hockey stick" nickname comes from how many option payoffs look: mostly flat (nothing much happens) until a point where the line suddenly bends and shoots away. That bend is the strike price doing its job. It's the moment the contract stops being a hopeful piece of paper and starts being a financial lever.

The axes matter, because they keep you from getting hypnotized by the shape. The horizontal axis (X-axis) is the *underlying price at expiration*. Think of it as the final scoreboard. Not today's price, not tomorrow's rumor—just the price on the date the contract settles. The vertical axis (Y-axis) is *profit/loss* for your position, usually measured in dollars per share (or per contract, depending on how you draw it). Above zero means profit. Below zero means loss. The whole picture is simply: "If the underlying ends *here*, my P/L ends *there*."

Once you accept that payoff diagrams are "expiration snapshots," they become liberating. They let you separate two questions that investors constantly mash together:

- *What do I believe will happen?* (a forecast problem)
- *What happens to me if I'm wrong?* (a design problem)

Most portfolio blow-ups aren't forecast failures. They're design failures—positions that punish you too much for being wrong at the wrong time.

Start with the simplest possible position: owning a stock (or an index fund, conceptually the same line). Suppose you buy at $100. At expiration, if the stock is $110, you're up $10. If it's $90, you're down $10. If it's $40, you're down $60. That's a straight line with slope 1: every $1 move in the stock becomes a $1 move in your P/L. Honest, clean, and emotionally brutal during crashes because there's no built-in stopping point. The line just keeps falling as the stock falls.

Now add the simplest "insurance overlay": a protective put. You still own the stock, but you also buy a put option with a strike—say $90—expiring on the same date. The put gives you the right to sell at $90 no matter how low the stock goes. In human terms, you've installed a floor under your position. You can still lose money, but below the strike the loss stops getting worse (beyond the premium you paid). The diagram tells you exactly what you bought: you didn't buy profit; you bought a *maximum bad day*.

There's a reason this picture is so powerful: it makes the trade-off visible. Above the strike, the put doesn't help. In calm or rising markets, the put behaves like an expense—your line is a little lower than "stock only" because you paid the premium. Below the strike, the put becomes the hero: the bleeding slows, then effectively caps. Without a diagram, people remember only the part they want to be true. With a diagram, both truths can sit on the page at the same time.

Right after you understand the axes, it helps to see the canonical diagram—the one you'll mentally reference every time someone pitches you "downside protection."

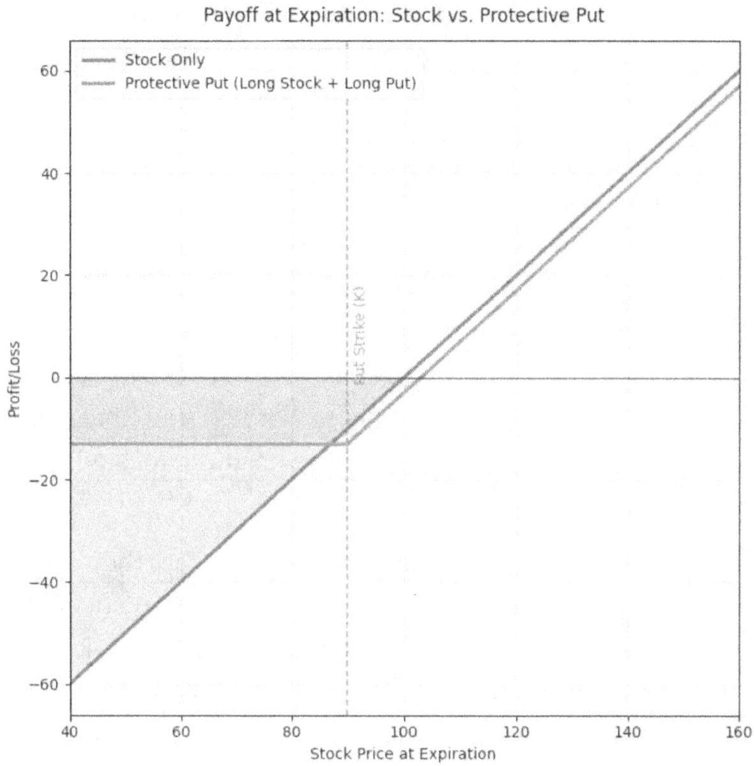

A diagram like this quietly answers questions that otherwise lead to late-night doomscrolling. If you're the kind of investor who says, "I just want to make sure I don't get wiped out," the protective put picture forces the follow-up: *Define 'wiped out.'* Is it a 10% drawdown? 20%? A level where you'd panic-sell? The strike is not a magical number. It's a choice about where you want your financial pain to change character.

Notice something else the chart reveals: the protective put doesn't eliminate loss. It *relocates* it. You take a small, known loss up front (the premium) to reduce the probability of a huge, unknown loss later. People who hate paying premiums usually aren't actually being "rational." They're reacting to a deeply human bias: certain small losses feel worse than uncertain big ones. Payoff diagrams help you fight that bias by making the uncertain big loss concrete.

The protective put is also a perfect example of why payoff diagrams are drawn "at expiration." Before expiration, an option's price can bounce around for reasons that feel unfair: volatility changes, time passes, markets gap at 9:30 a.m. The expiration payoff cuts through all of that and says: *Regardless of the drama, here is the end-state promise.* For hedging, that's often what you care about most—what the position does when the market forces you to make decisions.

That said, payoff diagrams have one built-in trap: they can lull you into thinking the hedge works like a light switch. "Below the strike I'm protected; above the strike I'm not." Real life is messier because you don't live only at expiration. You live through the path. In the middle of a crash, before the option expires, the put's market value can surge (which is helpful), but liquidity can be weird, spreads can widen, and your emotions can still do backflips. The diagram is still valuable—it's just not a full documentary of the experience. It's the contract's skeleton, not the market's weather.

Once you're comfortable reading the basic picture, you can start using payoff diagrams the way professionals use them: as a design tool. Want a tighter floor? Buy a higher-strike put. Want to spend less? Buy a lower-strike put and accept more pain before protection kicks in. Want protection for longer? Buy more time, usually at a higher premium. Every one of those choices has a geometric fingerprint. Your job is to decide which fingerprint matches your stomach.

It also helps to appreciate how the "stock only" line and the "protective put" line are telling two different stories about fragility. The stock-only line says: "My losses are unbounded on the downside." That's fragility—you're exposed to tail events, even if they're rare. The protective put line says: "My losses are bounded beyond a certain point." That's a form of robustness. You're paying to turn an unbounded problem into a bounded one. In portfolio terms, that can be the difference between rebalancing calmly and capitulating at the bottom.

There's an investing cliché that "risk is the price you pay for return." Payoff diagrams let you rewrite that into something more useful: *risk is the shape of your return distribution.* A protective put changes that shape. It doesn't make markets nicer. It makes *you* less breakable.

One more practical nuance: payoff diagrams are often drawn "per share," but real portfolios aren't one share; they're hundreds or thousands of shares, and options trade in standardized contract sizes (commonly representing 100 shares per contract for U.S. equity options). The diagram scales. The shape stays the same; the dollar amounts get louder. That's why sizing mistakes can be so punishing. A hedge that looks elegant on a whiteboard can become either meaningless (too small to matter) or financially suffocating (so expensive it drags returns) when applied to a real portfolio.

So when you look at a payoff chart, try reading it with two voices in your head. The first voice is the engineer: "Where are the kinks? What's the max loss? What's the slope?" The second voice is the human: "Would I still be able to sleep if the market opened down 7% tomorrow?" The entire point of hedging is to align those two voices.

Actionable takeaway: Pick one holding you care about (an index fund, a concentrated stock position, anything). On paper, draw a payoff diagram with two lines: "own it" and "own it + buy a put." Label the X-axis "Price at Expiration" and the Y-axis "Profit/Loss." Then circle the price level where you'd be most tempted to panic-sell. That circled point is a candidate strike zone for protection—and a reality check on whether you're trying to hedge risk, or hedge regret.

2.4 Futures and Linear Hedges: The Sledgehammer

If options are a seatbelt you only notice when something goes wrong, futures are the steering column itself. They don't wait for a crash. They move with the market, point for point, all day, every day. That's why they can feel refreshingly honest—and why they can feel merciless.

The core idea is simple: a futures contract is an agreement to buy or sell something later at a price set today. That "something" might be crude oil, corn, Treasury bonds, a stock index, or a currency. In modern markets you're usually not daydreaming about taking delivery of soybeans; you're using the contract as a risk dial. But the contract's DNA still comes from commerce: farmers and airlines and manufacturers have real-world budgets that get wrecked by unpredictable prices, and futures exist so those businesses can hand off that uncertainty to someone else.

That business origin matters because it explains why futures are built the way they are: standardized, exchange-traded, and ruthlessly mark-to-market. Futures markets aren't trying to be philosophical about risk. They are trying to keep promises enforceable at scale.

A futures hedge is called "linear" for the least exciting reason in finance: the math draws a straight line. If your futures exposure is equivalent to one unit of the underlying, then a $1 move in the underlying produces about a $1 change in your futures position at settlement. No bend. No floor. No "it only kicks in below a strike." Futures are the opposite of the hockey stick. They're the ruler.

That straight-line behavior is exactly why futures are described as a sledgehammer. Sometimes you don't need a delicate tool. Sometimes you need to reduce risk right now, with high precision, without selling your long-term holdings, triggering taxes, or ripping up your allocation plan. A linear hedge does that. It doesn't negotiate; it offsets.

Here's the contrast in plain language. A protective put changes the *shape* of your downside: you pay a premium for the privilege of being less fragile in a crash. A futures hedge changes your *amount* of exposure: you decide how much market you want to carry, and you carry less. One is insurance. The other is exposure control.

Suppose you own a broad equity portfolio and you're staring at a near-term risk-maybe you have a large cash need soon, or maybe you simply can't stomach the next few months of volatility without making a bad decision. You could sell some shares. That would certainly reduce risk. But it would also change your portfolio permanently unless you buy back in, and it may create taxes or timing risk. A futures short on a stock index is like putting a temporary counterweight on the portfolio. Your stocks stay in the garage; you're just putting a weight on the other end of the seesaw.

The cleanest way to see futures is to imagine a forward contract in a backyard setting. Two neighbors agree today that in 90 days one will buy the other's used car for $10,000. If the market for used cars collapses, the buyer overpays relative to market—bad for the buyer, good for the seller. If used-car prices surge, the buyer gets a bargain—good for the buyer, bad for the seller. That's a forward. A futures contract is the exchange-traded, standardized, highly engineered cousin of that simple agreement.

The elegance is that you don't have to own the asset to hedge it, and you don't have to sell the asset you do own. You just take a position that moves against it. If you're long stocks and you short stock-index futures, you've created a partial cancellation. When stocks drop, your portfolio hurts—but the short futures position tends to make money, offsetting the pain. When stocks rise, your portfolio gains—but the short futures position loses, giving back some upside. Again: no magic, just subtraction.

That "give back some upside" part is where futures feel emotionally different from options. With a put, you usually keep your upside (minus the premium). With a futures hedge, you *deliberately* dull your upside for as long as the hedge is on. Futures don't wait for disaster; they tax every rally. That's why they're often used as temporary risk reducers, not as permanent portfolio accessories.

If futures are so straightforward, why don't more long-term investors use them? The answer is not that futures are "too advanced." It's that futures force you to confront the most adult concept in markets: *margin*.

Margin in futures is widely misunderstood because the word is reused in different contexts. Futures margin is not a down payment on the asset. It's a *performance bond*. It's money posted to prove you can keep your promises as prices move. You're not "buying" $200,000 of stock exposure by putting down $10,000 like it's a layaway plan. You're entering a contract with $200,000 of notional exposure, and the exchange requires you to post collateral to reduce the odds you default.

The second mechanical feature is what makes futures so effective and so unforgiving: they are marked-to-market. That means gains and losses are recognized *daily*. If the market moves against you, your account is debited. If it moves in your favor, it's credited. There's no pretending. There's no "I'll just wait until expiration and it'll come back." The contract updates your reality every day.

That daily settlement is why futures can create a very specific kind of stress: not "am I right over a year?" but "do I have cash today?" In hedging, cash-flow risk can be more dangerous than price risk. A perfect hedge that forces you to liquidate at the worst moment because you can't meet a margin call is not a hedge—it's a trap with good intentions.

So the way to respect futures is to think in two layers at once:

- *Exposure layer:* how much market risk you're offsetting (the notional hedge size).
- *Cash-flow layer:* how much collateral you may need to post if the market moves against your hedge.

You can get the exposure layer "right" and still get wrecked by the cash-flow layer. Long-term investors are often surprised by this because their default investing tool—index funds—doesn't come with a daily cash demand.

The phrase "linear hedge" also hides a subtle portfolio truth: futures are usually used to hedge a *bucket* of risk, not a specific security. If you own 40 individual stocks, shorting a broad index future won't perfectly offset each name. The hedge is on the market factor, not the idiosyncratic factor. That's often exactly what you want. In a crash, correlations tend to rise, and "market risk" becomes the bully in the room. A broad futures hedge targets the bully. It won't protect you from a single stock scandal, but it can protect you from the entire room tipping over.

Another reason futures can be the right tool is cost transparency. With options, the cost is an upfront premium that can feel like a recurring insurance bill. With futures, there's no premium in the same way. You're not paying an explicit check to hold the contract (though there are implicit costs like financing, roll, bid-ask spreads, and the opportunity cost of posted margin). The absence of a premium can make futures feel "free," which is emotionally pleasant and financially dangerous. If you hedge with futures, the cost often shows up as foregone upside during rallies and potentially painful cash calls when markets move against your hedge.

A good mental model is this: options are a way to buy a *conditional* offset. Futures are a way to install an *unconditional* offset. Conditional offsets are expensive but patient; unconditional offsets are efficient but relentless.

There's also a behavioral angle that's worth saying out loud. Many investors think they want protection, but what they really want is protection that doesn't interfere with their optimism. A protective put is easy to love in bull markets because the stock still climbs; the hedge just quietly costs money. A futures hedge is harder to love because it's visibly "in the way." The market goes up and your hedge loses money, right in front of you, like a critic with a megaphone. That visibility can cause investors to abandon the hedge at the worst possible time—after a rally, right before the next drop—turning risk management into a performance of bad timing.

Futures also introduce a practical question that options often avoid: *What does "hedged" actually mean?* If you hedge 100% of your equity exposure with futures, you've essentially converted yourself into a cash-like investor for the duration of the hedge (minus basis effects and imperfect tracking). That might be exactly what you need for a known liability. It might also be a self-inflicted opportunity cost if you do it out of fear and then freeze. More commonly, investors use futures to hedge a portion—enough to make the portfolio psychologically survivable without turning it into a different life plan.

If the sledgehammer metaphor feels a little harsh, it's because linear hedges don't pretend to be gentle. They are best used when you have a clear objective that can be expressed as a number: "Reduce equity exposure by roughly 30% for the next quarter," or "Bring portfolio volatility down enough that I won't break the plan." Options are better when your objective is more like a boundary: "I can accept the first 10% down, but not the next 20%."

That difference—*dial* versus *floor*—is the simplest way to choose between the two tools without drowning in jargon.

One last mechanical note that keeps futures from feeling mystical: you usually don't "hold to delivery." Most futures positions are closed by taking the opposite position before expiration. Long one contract? Sell one contract later to flatten. Short one contract? Buy one contract later to flatten. This is how futures become a portfolio tool rather than a warehouse problem. The exchange standardization makes it possible: you're trading interchangeable contracts, not negotiating bespoke deals with a neighbor.

Used carefully, futures can be a remarkably clean way to manage risk. They're not primarily for predicting where markets go; they're for controlling what markets *do to you* while you keep your long-term portfolio intact. Used carelessly, they turn into leveraged bets that demand cash at the worst possible moment. The difference is almost never intelligence. It's preparation, sizing, and respect for margin mechanics.

Actionable takeaway: Before using any futures hedge, write down two numbers: (1) the percentage of your portfolio you want to neutralize (your exposure objective), and (2) the maximum cash you are willing and able to post on short notice (your margin-stress budget). If you can't comfortably support the second number, don't use a linear hedge—choose a tool with defined, prepaid downside instead.

2.5 ETFs and Inverse Products: Handle with Care

The most expensive word in investing is "simple." Inverse ETFs are often sold—sometimes literally, sometimes just in the way people talk about them—as a simple hedge: "If the market goes down, this goes up." That sentence is true in the narrowest possible sense and dangerously incomplete everywhere else. Inverse products are not insurance policies, and they are not a free shortcut around learning options or futures. They are a specific kind of machine that resets itself every day, and that daily reset quietly rewrites your returns.

The appeal is obvious. You can buy an inverse ETF in the same brokerage account where you buy an index fund. No approvals. No options chain. No margin discussion that makes your pulse quicken. It feels like hedging with training wheels. And that's exactly the trap: the interface is familiar, so people assume the math is familiar.

An inverse ETF typically targets the opposite of an index's *daily* return. If the index is down 1% today, a -1x inverse ETF aims to be up about 1% today (before fees and tracking noise). Leveraged inverse ETFs (like -2x or -3x) aim for a multiple of the opposite daily move. The word to tattoo on your brain is *daily*. The target is not "-1x for the next year." It's "-1x for today, then reset, then do it again tomorrow."

That reset is not a marketing quirk—it's the core mechanism. The fund rebalances so that the next day it again targets -1x (or -2x, -3x) exposure to whatever its reference index is doing that day. The result is path dependence: what you earn over weeks or months depends not just on where the index starts and ends, but on how it wiggles along the way. Wiggles cost money.

To see why, you don't need calculus. You need one annoying truth about losses: percentages don't cancel cleanly.

If an index drops 10%, you're at 90. If it then rises 10%, you're at 99, not back to 100. Nothing scandalous happened; it's just arithmetic. The drop was applied to a bigger number than the rebound. This is the "math of loss," and it's the soil where inverse ETF decay grows.

Now layer in the daily reset. An inverse ETF is effectively compounding daily inverse returns. In a volatile, back-and-forth market that ends roughly flat, the inverse ETF can bleed value even though your intuition screams it should "also end flat." The intuition assumes a clean mirror image. The product delivers a daily mirror image that compounds into something else.

Think of it like walking up a down escalator. If the escalator is moving steadily down and you walk steadily up, you make progress. But if the escalator speeds up and slows down unpredictably while you keep adjusting your pace to "match" it each day, you waste energy. The daily reset forces constant adjustment. The adjustment is where returns leak.

The leak has a few names: *volatility decay, beta slippage, compounding drag*. You can call it whatever you want. The lived experience is simple: holding inverse ETFs for longer than their intended horizon can turn a hedge into a slow-motion loss.

It gets worse when investors use inverse ETFs the way they use insurance: buy it and forget it. Insurance is designed to be held. A protective put has an expiration date, yes, but it's built to behave like a floor over a defined period. A futures hedge is linear and demands margin, but at least its relationship to the underlying is straightforward. An inverse ETF is neither. It's a daily rebalanced exposure wrapper. It is a *trading* instrument wearing an *investing* costume.

Here's the visual that should permanently change how you think about these products. Imagine an index that thrashes around—up, down, up, down—yet ends the period basically where it started. If "inverse" meant "mirror," the inverse ETF should also end basically where it started. But with daily compounding, it can drift lower and lower, like a bar of soap slowly dissolving in the shower.

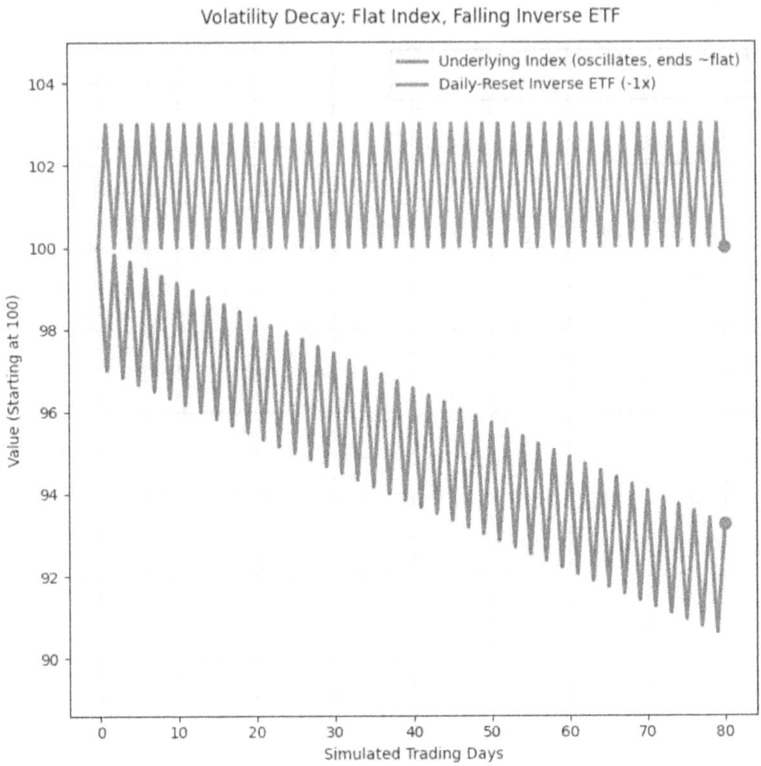

Stare at that for a moment. The index is basically doing jumping jacks in place. The inverse ETF is quietly losing weight. Nothing "went wrong" with the product. It did what it said: it targeted the opposite of the index's *daily* move. Your brain just wanted it to do something else: deliver the opposite of the index's *multi-day* move. That's a different contract.

Now add real-world frictions and the story becomes even less forgiving. Inverse ETFs charge fees. They incur transaction costs to rebalance. They often gain exposure using derivatives (like swaps and futures), which can introduce additional tracking differences. In quiet markets these frictions are small.

In choppy markets—the exact markets where people reach for a "hedge"—those frictions have more opportunities to bite.

Leveraged inverse ETFs are the extreme version of the same idea. A -2x or -3x product is like turning up the volume on both the intended effect and the unintended decay. The daily compounding becomes more intense. The divergence between what you *expected* ("two times the opposite!") and what you *get* over time can become jaw-dropping.

There's also a temptation to treat inverse ETFs as a clever substitute for selling. "I don't want to reduce my equity exposure; I'll just buy some inverse." That can work as a short-term tactical patch. But for longer holding periods it creates a strange portfolio: you own a long-term compounding asset (stocks) and pair it with a short-term compounding trading instrument that is structurally hostile to time. It's like bolting a parachute to your car and leaving it deployed on the highway. Sure, you're less likely to speed. You're also constantly dragging fabric.

Another subtle hazard is psychological. Because inverse ETFs trade like stocks, they invite the same casual behavior: buy, hold, ignore. But their risk profile demands the opposite behavior: monitor, rebalance, define holding periods, and accept that the hedge itself can be a source of loss even if the market doesn't trend. Many investors are not set up—emotionally or operationally—to do that. A hedge that requires daily attention is not a hedge for most people; it's a second job.

It's worth drawing a sharp line between "inverse ETF as a short-term tactical tool" and "inverse ETF as a long-term safety net." The first can be defensible if you understand the daily objective and you're honest about the time window. The second is where portfolios go to die by a thousand paper cuts. Regulators have repeatedly warned that many leveraged and inverse products are designed to meet their objectives on a daily basis, and that holding them longer can produce results that surprise investors—especially in volatile markets. The surprise is not subtle. It's structural.

ETNs (exchange-traded notes) add another wrinkle: credit risk. An ETN is essentially unsecured debt issued by a bank or financial institution, with returns linked to an index. That means you're not just taking market exposure; you're taking issuer exposure. Most of the time that feels like a footnote. In the moments when credit risk suddenly matters, footnotes have a nasty habit of becoming headlines.

So what should a long-term investor do with all this?

First, don't confuse convenience with safety. The fact that you can buy an inverse ETF with one click doesn't mean it behaves like a sensible long-term hedge. Many people reach for inverse ETFs because they want the emotional relief of "doing something" without learning options mechanics or futures margin. That's understandable. It's also a recipe for buying the wrong tool because it has the right user interface.

Second, get extremely specific about your *time horizon*. If you are using an inverse ETF, you are implicitly saying: "My hedge objective is daily, and I am prepared to manage it as a daily instrument." If that sentence makes you tired, listen to the fatigue. Your future self is trying to protect you.

Third, recognize what inverse ETFs are actually good at: expressing a short-term bearish view or temporarily reducing exposure when you have a defined, short window and you can monitor the position. They are not a set-it-and-forget-it crash policy. They are not a substitute for a thoughtfully sized put structure. And they are not a free lunch.

None of this is meant to shame anyone for wanting an accessible hedge. It's meant to prevent a particular heartbreak: watching a "protective" position lose money month after month in a market that feels like it isn't even going anywhere. That experience doesn't just hurt returns; it corrodes trust. It makes people conclude that hedging is pointless, when the real problem was that they bought a daily instrument to solve a long-term problem.

Actionable takeaway: If you're considering an inverse ETF, write its goal in one line using the word *daily*. For example: "This product targets -1x the *daily* return of [index]." Then write your intended holding period in calendar days. If your holding period is more than a few days and you're not planning to monitor and adjust, don't use it as a hedge—use a tool whose payoff is designed to match your time horizon (like prepaid option protection) or reduce exposure directly.

2.6 Volatility as an Asset Class

There's a particular moment in every bear market when someone who has never talked about volatility in their life suddenly becomes a volatility philosopher. It's usually after a brutal down day, when your portfolio feels like it's been dropped on concrete and the financial news has the tone of a disaster movie trailer. That's when you hear the seductive idea: *What if I could just buy fear? If markets panic, I profit. Problem solved.*

That fantasy has a name: "long volatility." And it's rooted in a real observation. When stocks fall hard, measures of expected volatility tend to spike. Fear shows up as higher option prices, and higher option prices show up as a higher volatility index. So the instinct makes sense: if fear rises when your stocks fall, why not own fear?

Because "fear" is not an asset. It's a condition. And conditions are expensive to warehouse.

The poster child is the VIX, often called the market's "fear gauge." People talk about it the way they talk about oil or gold, as if it's a thing you can buy and put in a vault. It isn't. The VIX is an index number derived from option prices on the S&P 500 (more precisely, it's built from a strip of S&P 500 index option quotes across strikes, targeting a constant 30-day horizon using options with expirations roughly between 23 and 37 days). In plain English: the VIX is the market's *implied* expectation of how violently the S&P 500 might swing over the next month, backed out of the price of insurance. It's a thermometer reading, not the weather.

You can't buy a thermometer reading and expect it to keep you warm.

This is the first trap in "volatility as an asset class": confusing an index with a tradable instrument. With stocks, the index and the thing you buy are close cousins. The S&P 500 is an index, but you can buy an ETF that holds the underlying stocks and tracks it reasonably well. With volatility, the index and the tradable products are more like distant relatives who don't return each other's calls.

Tradable volatility exposure usually comes through derivatives on volatility-most notably VIX futures and VIX options-and through exchange-traded products (ETFs/ETNs) that hold or roll those futures. That "futures" word is not a footnote. It's the whole story. When someone says, "I bought VIX," what they typically bought is a product linked to VIX *futures*, not the spot VIX itself. The difference is where the long-term bleeding begins.

To understand why, remember the two shapes of hedging tools. Options can create convex payoffs: you pay a known premium and you get a payoff that accelerates in stress. Futures are linear: efficient, but they require constant management and they settle daily. Volatility products sit in a third category: they are often a *rolling position* in futures contracts. You're not buying a payoff at a single expiration date. You're buying a process.

And processes have carrying costs.

If you own a VIX futures-based product for more than a quick tactical window, it typically has to "roll" its exposure: as the front-month futures contract approaches expiration, the product sells it and buys a later-dated one. This sounds like housekeeping. In practice, it's where your returns get taxed by market structure.

Here's the intuition without the jargon. Imagine you want ongoing insurance. You don't buy one policy forever; you keep renewing it. If renewal prices are usually higher than the expiring policy—because the market is calmer now and expects volatility to rise back toward normal later—then you're constantly selling cheaper insurance and buying more expensive insurance. That repeated "sell low, buy high" is a structural drag. In volatility-land it's called *roll yield*. In human-land it's called *you are paying rent to live in fear*.

In many normal environments, the VIX futures curve is upward sloping: longer-dated futures trade at higher prices than near-dated futures. That shape is often called contango. You don't need to memorize the term. You just need to recognize the consequence: a long volatility product that has to roll forward is often buying more expensive contracts than the ones it's selling. That's a headwind you feel even when markets do nothing dramatic.

This is why so many volatility ETPs look like tragic long-term charts: occasional spikes up during crises, followed by long stretches of decay. The spikes are real-volatility can explode when the market panics. But the decay is also real, and it's not bad management or "tracking error" in the casual sense. It's the price of maintaining a position in a mean-reverting, insurance-like quantity through a rolling futures mechanism.

Mean reversion is the second big trap. Volatility is not like earnings growth. It doesn't compound upward over decades. Volatility tends to surge and then fall back. It behaves more like heart rate than like net worth. If you buy volatility after it spikes, you are often buying it at the exact moment it is statistically more likely to decline. That doesn't mean it *must* decline immediately—crises can persist—but it means "buy and hold forever" is fighting the natural behavior of the thing you're holding.

There's also a subtle but important difference between *expected* volatility and *realized* volatility. VIX is derived from option prices, which embed not just expectations but also a premium for bearing risk-the price of insurance, not merely the forecast of storms. This is why "volatility" can be expensive even if the market ultimately doesn't swing that much. You're not buying a neutral prediction; you're buying something closer to a priced fear contract.

If that still feels abstract, consider how fear is manufactured in real time. The VIX is quote-based. It is built from options quotes, and in stressed markets those quotes can widen as liquidity thins and dealers protect themselves. During sudden shocks, the microstructure of the options market-who is quoting, how wide the bid-ask spreads are, how aggressively dealers mark prices-can mechanically amplify volatility readings. That doesn't make the VIX "fake." It makes it a real-time indicator built from a real market that is itself reacting to stress. In other words: the fear gauge can jump not only because investors are afraid, but because the plumbing of trading under stress has its own reflexes.

This matters for investors because it highlights how different volatility exposure is from owning stocks. When you buy an index fund, you're buying claims on future cash flows. When you buy volatility exposure through futures-based products, you're buying something that is closer to a continuously repriced insurance strip-sensitive to market mood, market mechanics, and the cost of rolling contracts. It's not "more sophisticated." It's just a different animal, and it lives on different food.

So is "volatility as an asset class" useless? Not at all. It can be powerful in the right role. The question is: what role are you trying to fill?

If your goal is *crash response*-something that tends to pop when equities gap down-long volatility exposure can help. It often behaves like a fire extinguisher: dormant most days, suddenly valuable when there's smoke. The problem is that fire extinguishers have a shelf life and a carrying cost if you insist on holding them in your hands all the time. Long volatility is rarely a "core holding." It's more like a tool you deploy when the probability-weighted benefit justifies the drag.

If your goal is *drawdown control* over a defined window, volatility exposure can sometimes be part of the solution-but it's usually a complement, not a replacement, for clearer hedges like puts. Puts give you a contractually defined relationship to a price level. Volatility products give you exposure to the market's expectation and pricing of future swings. Those are related, but they're not the same. In a fast crash, volatility can spike and help. In a slow grind down, volatility might not behave the way you emotionally expect. In a whipsaw market, the decay can be punishing. Buying "fear" does not guarantee you make money when you feel fear.

If your goal is *long-term portfolio insurance*, volatility products are usually the wrong wrapper. The structural costs—roll yield, fees, compounding effects—tend to work against you over time. You may still choose them for a tactical hedge, but you should do it with eyes open and with a defined exit plan.

A practical way to keep your head straight is to stop asking, "Will volatility go up?" and start asking, "What exactly is my instrument holding, and what does it cost to keep holding it?" With volatility ETPs, you are often paying for:

- Exposure that must be rolled forward repeatedly (structural drag in many environments)
- Exposure that is tied to futures, not spot VIX (basis risk)
- A quantity that mean-reverts (timing risk)
- A market that can behave oddly under stress because it's quote-driven (microstructure risk)

You don't have to fear those risks. You just have to stop pretending they don't exist.

There's also a common mistake that sounds prudent: "I'll just buy a little volatility product and leave it there as a permanent hedge." The issue is that "a little" is rarely stable. In calm markets, the position may decay and become *too little* to matter. In a spike, it may suddenly become *a lot*, and you may not rebalance because the gain feels like vindication. Over time, your "little hedge" can drift into a weird, unmanaged side portfolio with its own personality. A hedge that you don't rebalance is not a hedge; it's a bet that changes size by accident.

None of this is meant to discourage curiosity. Volatility is fascinating precisely because it's the market's nervous system made visible. It's also a reminder that the market prices not just outcomes, but the discomfort of uncertainty. When you buy volatility exposure, you are paying for that discomfort in a very literal way.

So the right attitude toward volatility products is the same right attitude toward any powerful tool: respect the mechanism, define the mission, and set a time horizon that matches the contract's economics.

Actionable takeaway: Before buying any volatility-linked ET-F/ETN, look up what it actually holds (spot index, VIX futures, a rolling basket, etc.) and write down a calendar exit date *before* you place the trade. If you can't explain, in one sentence, how the product maintains exposure (especially whether it rolls futures), don't buy it as a hedge. "Buying fear" only works when you understand what you're actually holding-and what it costs to keep holding it.

Chapter 3

The Price of Insurance: Valuing Protection

If you could buy car insurance only while your vehicle was already skidding on ice, you would expect to pay a massive premium. Financial markets work similarly-protection is most expensive exactly when you feel the greatest need for it. In this chapter, we will decode the price tag of safety. We will move beyond complex formulas to understand the intuition behind what options cost, how market fear drives those prices, and how to spot a 'bad deal' before you buy it.

3.1 Drivers of Option Pricing

The first time you buy a put for protection, the premium can feel insulting. You're not buying a share. You're not buying a bond that sends you interest. You're buying a *right*: the right to sell at a certain price (the strike) before a certain date (expiration). The market hands you a receipt for that right and charges you for it upfront. That receipt is the option's *premium*, and it's best understood the way you'd understand any other insurance bill: part of it pays for something real *today*, and part of it pays for uncertainty you're reserving for *later*.

There are only two ingredients in that bill:

- *Intrinsic value*: what the option would be worth if it expired right now.
- *Extrinsic value* (often called *time value*): everything else you're paying for—time, uncertainty, and the possibility that the world gets weird before the clock runs out.

Once that clicks, option pricing stops feeling like a magic trick and starts feeling like a quote from an insurer. Change the amount of time on the policy, or change the perceived risk of a claim, and the quote changes—sometimes violently.

A quick concrete anchor helps. Imagine a stock at $100 and a put with a $105 strike. You have the right to sell at $105 even though the market is at $100. That right is immediately useful. If the option expired this second, you could (conceptually) sell at $105 and buy back at $100 for a $5 gain. That $5 is intrinsic value. Now imagine a $95 strike put on that same $100 stock. If it expired this second, you wouldn't use it; you can already sell at $100 in the market, so a $95 sell-right is worthless *today*. Its intrinsic value is $0. Yet it still costs money, because the next few weeks could turn that currently-useless right into a lifesaver.

Right after you internalize that split, you'll notice something slightly uncomfortable: most of the time, when you buy protection, you're paying mostly for extrinsic value. You are buying the *possibility* of needing it. That is the nature of insurance. If you want a hedge that's cheap, you usually have to accept that it's also far away from being useful today.

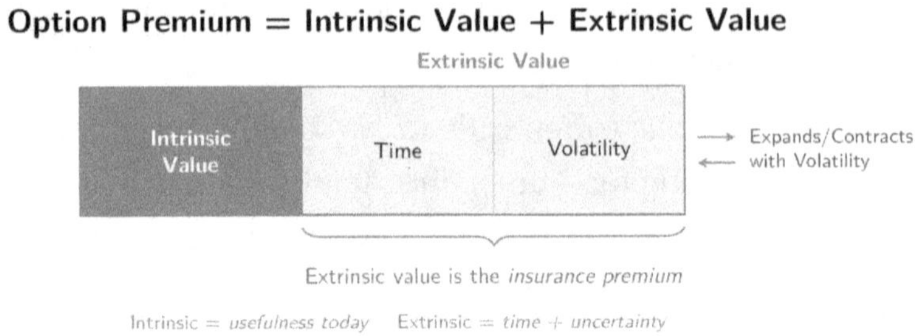

Intrinsic value is the blunt, mechanical part. It ignores stories, headlines, and drama. It only asks: *If we froze time, would this right be worth exercising?* For a put, intrinsic value is essentially "how far the stock is below the strike," but never less than zero. If your put's strike is above the current price, the put is in-the-money and has intrinsic value. If the strike is below the current price, the put is out-of-the-money and has no intrinsic value.

Extrinsic value is the part with personality. It's the market admitting it doesn't know the future—and charging you for that humility. An out-of-the-money put can go from "worthless today" to "painfully valuable" in one ugly week. Extrinsic value is what you pay for the chance to be right *later*, even if you're wrong *now*.

If intrinsic value is the option's *what*, extrinsic value is the option's *why*. And the two levers that move extrinsic value more than anything else are:

- *Time to expiration*: how long the option has to become useful.
- *Implied volatility*: how wild the market believes the ride could be.

Interest rates and dividends matter too, but for most long-term investors using listed equity and index options as a hedge, time and volatility are the knobs you'll feel in your bones.

Time: the meter that never stops running

Time value is like paying rent on a fire extinguisher. You don't pay because you expect a kitchen fire on Tuesday. You pay because you want the extinguisher in the cabinet for the months when life happens. An option with more time has more ways to surprise you—and surprises are what make options worth owning.

Time does something counterintuitive: it is both what you're buying and what is being taken away from you. Every day that passes is one less day for the market to deliver the move your hedge needs. That daily erosion is called *theta decay*. The phrase sounds technical, but the experience is not. You check your hedge a week later, nothing terrible happened, and the option is cheaper. That is theta doing its quiet work.

The cruel part is that theta doesn't drain like a battery with a steady percentage. It behaves more like an hourglass: the sand falls faster as you approach the end. Short-dated options can feel like they *evaporate*. Long-dated options decay more slowly at first, which is why many long-term hedgers prefer paying more upfront for longer expirations: it spreads the "rental cost" over more calendar days and reduces the pressure to constantly replace expiring protection.

But longer time isn't free; it's simply a different form of expensive. When you buy more months, you're buying more uncertainty, and the seller demands to be paid for carrying that risk.

A 12-month policy costs more than a 1-month policy even if the house is the same. That's not greed; that's arithmetic.

There's a practical way to think about time without formulas: ask how many *decision points* are left. A one-week option has only a few trading days to be rescued by a selloff. A one-year option has earnings seasons, policy meetings, election cycles, and "something broke in the plumbing of the financial system" moments. More time means more possible pathways into trouble. The market prices those pathways.

Volatility: the thermostat on insurance prices

Volatility is how much prices swing—the market's speed and violence, not its direction. Two markets can end the year at the same level, yet feel completely different to live through. One can glide. The other can whip-saw your emotions daily. Options care deeply about that difference.

Implied volatility (IV) is the volatility level that is embedded in option prices—the market's best collective guess (with a profit margin) about how bumpy the road ahead might be. When IV rises, option premiums generally rise, even if the stock price hasn't moved much yet. That surprises many investors the first time they watch it happen. They expect the hedge to get expensive *after* the drop. Often it gets expensive *before* the drop, because fear shows up in pricing the moment people start reaching for insurance.

Here's a feel-it-in-your-wallet example. Suppose two identical stocks are at $100. One has been trading like a calm lake; the other has been trading like a caffeinated squirrel. A $95 strike put with 60 days to expiration is the same contract structure in both cases—same strike, same time. Yet the "squirrel stock" put will almost always cost more, because there's a higher chance the stock ends up below $95 before expiration. The option is a claim on *movement,* and movement is more likely when volatility is higher.

That's why volatility is often described as the market's fear gauge. Fear doesn't just mean "down." It means "unpredictable." Options are priced on the size of possible outcomes, not on comforting narratives about long-term averages.

There's another twist that matters for hedging: implied volatility tends to rise when stocks fall. In a selloff, the stock moves against you *and* the price of insurance often jumps at the same time. This is great when you already own the protection (your put can gain value from both the underlying decline and the volatility spike). It's terrible when you're trying to initiate or roll the hedge in the middle of the panic (you're buying the extinguisher while the kitchen is already on fire).

Moneyness: why "distance to the strike" changes the mix

Even though time and volatility are the levers, the option's relationship to the current price—its *moneyness*—determines what you're actually paying for in practice.

An at-the-money option (strike near the current price) is often dominated by extrinsic value. It doesn't have much intrinsic value because it's not deeply in-the-money, but it has a lot of *possibility.* It could end up meaningfully in-the-money with a modest move. That possibility is expensive.

A deep in-the-money put has a lot of intrinsic value. You're paying for a right that is already useful. It often behaves more like a substitute for selling the stock (or shorting), and less like a lottery ticket on a crash. For a hedger, deep ITM can mean more immediate coverage but a bigger cash outlay.

A far out-of-the-money put (the classic "crash insurance" strike) usually has zero intrinsic value and is almost entirely extrinsic. That can be psychologically appealing—"I'm only risking the premium"—but it can also be deceptively hard to use well. You need a large, fast drop to turn it from potential into intrinsic value before the clock runs out. The premium is smaller, but the probability of payout is lower. Cheap insurance often means narrow coverage.

This is where many hedging programs quietly fail: not because puts "don't work," but because the investor bought a strike so far away that the hedge only pays in an apocalypse, then feels betrayed during an ordinary bear market. The contract did exactly what it was designed to do. The expectation was the problem.

A useful mental model: intrinsic is a floor, extrinsic is a mood

Intrinsic value is a floor: it's what you'd keep if you squeezed the option down to the moment. Extrinsic is a mood: it expands when people imagine chaos and shrinks when people get complacent.

That mood can change faster than any other part of a portfolio. A stock can drift 1% in a day, but implied volatility can jump 30% in an afternoon because a central bank chair cleared their throat in a way the market didn't like. Options translate that shift in mood into dollars immediately.

This is why hedges can feel "unfair" when you're shopping for them. When things are calm and you finally feel rational enough to buy protection, premiums can look relatively reasonable—but you also feel least motivated to pull the trigger. When you feel most motivated, premiums are often highest. The market is not trying to be kind; it's trying to clear supply and demand for fear.

For a long-term investor, the aim isn't to outsmart this mood swing every time. The aim is to recognize which part of the premium you're paying for and whether it matches the job you need done. If your primary worry is a near-term shock, you're buying time and volatility sensitivity. If your primary worry is an eventual drawdown at some unknown point, you're buying a longer reservation on uncertainty, and you should expect to pay for that reservation.

How time and volatility interact: the "longer lease, higher thermostat" effect

Time and volatility don't just add to each other; they amplify each other.

Longer time increases the chance that volatility has an opportunity to matter. If you buy a one-week option, even high implied volatility has limited time to deliver a huge move. If you buy a one-year option, volatility has many more chances to do damage—or to do you a favor if you're hedged.

That's why an increase in implied volatility tends to raise the price of longer-dated options by more dollars than the same volatility increase raises a very short-dated option (all else equal). Think of it like this: if the insurer thinks the neighborhood is getting riskier, the price increase hurts more on the 12-month policy than on the 7-day policy.

In real portfolios, this interaction shows up as a tradeoff between *roll stress* and *upfront cost*. Short-dated hedges are cheaper per contract but force you into frequent renewals, which means you're repeatedly exposed to the market's mood when you re-buy. Longer-dated hedges cost more per contract but reduce the number of times you must go shopping during a panic.

Why two "similar" puts can have wildly different prices

If you've ever compared two puts and thought, "These look basically the same; why is this one twice the price?" the answer is usually hiding in one of these places:

- *Time*: even a few extra weeks can meaningfully increase extrinsic value.
- *Implied volatility*: the market may be pricing a storm you haven't fully noticed.
- *Moneyness*: one put might already be doing real work (intrinsic), while the other is mostly hope (extrinsic).

Here's an intuitive exercise that keeps you honest: pretend the option expires *today*. What would it be worth? Whatever number you get is intrinsic. Everything above that is the price of waiting plus the price of uncertainty. If the "above that" part feels shockingly large, you're not seeing a conspiracy; you're seeing a market that is charging a lot for fear and time.

This framing also helps with a common emotional mistake: celebrating when a hedge is "cheap" without asking why. If implied volatility is low, premiums often look attractive. That can be a good time to buy. It can also be a time when markets are relaxed for a reason—steady growth, strong liquidity, no immediate catalysts. Low IV is not a coupon; it's a diagnosis. Sometimes it's a diagnosis worth acting on. Sometimes it's simply the market being calm.

Actionable takeaway: decompose every premium before you buy

Before entering any hedging trade, force a one-minute decomposition:

- Compute (or approximate) *intrinsic value*: what the option is worth if it expired now.
- Label the rest as *extrinsic value*: what you're paying for time and implied volatility.
- Ask one blunt question: *Am I comfortable paying this much for possibility?*

If most of what you're paying is extrinsic (it often will be), don't treat the premium like a "cost" in the abstract. Treat it like rent. Then decide whether you want a shorter lease (cheaper, higher roll pressure) or a longer lease (more expensive, less roll stress), and whether the current volatility thermostat makes that rent feel like a fair price for the protection you're trying to reserve.

3.2 Implied vs. Realized Volatility

Buy a hedge on a quiet Tuesday and you'll discover an uncomfortable truth: the market is willing to sell you disaster insurance, but it refuses to do it at cost. The price includes a markup for the unpleasant experience of being the one on the other side. That markup is the *fear premium*, and it's one of the main reasons long-term hedging can feel like paying for an alarm system that mostly chirps at you for low battery.

Volatility is the language that markup speaks. Not the volatility you can measure after the fact, with a spreadsheet and a rearview mirror, but the volatility the market *implies* through what people are willing to pay right now.

Two volatilities, two different jobs

Realized volatility (also called historical volatility) is what actually happened. It's measured from past price moves: how jumpy the market *was*. If you look back over the last month and the S&P 500 barely wiggled, realized volatility will be low. If it careened around like a shopping cart with one bad wheel, realized volatility will be high.

Implied volatility is what's embedded in today's option prices. It's the market's "best guess," but that phrasing is almost too polite. Implied volatility is a clearing price for anxiety. It answers a very specific question: *What level of future wiggle would make this option price make sense in a pricing model?* In other words, IV is the volatility number you must plug in to get the model to match the quote. The market doesn't publish its reasoning in a neat memo; it publishes it in the premium.

That distinction sounds academic until you're actually hedging. Realized volatility tells you what kind of storm you *just* lived through. Implied volatility tells you how expensive umbrellas are *today*.

Why implied volatility so often runs hot

If option prices simply reflected the average future realized volatility, protection would be "fair" in the way homeowners insurance might be fair in a world with perfectly predictable house fires. But the market isn't a charity and the future isn't a stable lab experiment. Options are traded in crowds, under stress, with real money and real career risk.

So implied volatility tends to sit above subsequent realized volatility for long stretches of time. The simplest explanation is also the most human: lots of investors want to buy protection, especially when they're nervous, and fewer investors are naturally eager to sell it. The sellers who *are* willing to underwrite crash risk demand compensation for doing something that feels like picking up pennies in front of a truck.

That persistent gap is often called the *volatility risk premium*. Don't let the name intimidate you. It's just the idea that "insurance" tends to be priced so the insurer has a positive expected return. Sometimes the premium is small and you barely feel it. Sometimes it's enormous and you feel like you're paying for a policy written by a pessimist with a yacht payment due.

One more reason IV runs hot: volatility isn't evenly distributed through time. Markets spend many days drifting and a few days falling down the stairs. Realized volatility averages those days together. Option sellers, meanwhile, worry about the staircase days because that's when they can lose a lot, fast. They price for the tail events. The buyer pays for the *shape* of risk, not just the average.

A hedger's paradox: you want to buy when it feels unnecessary

When implied volatility is low, the market is relaxed. Premiums can look "cheap." That's also when buying protection feels hardest, because it can feel like paying for something you don't need. When implied volatility is high, the market is tense. Premiums look painful. That's when buying protection feels most urgent, because the risk is loud.

This is the emotional trap that makes many hedging programs inconsistent. Investors procrastinate during calm periods, then attempt to buy insurance during the fire drill. But option markets are not slow-moving. They reprice fear quickly. If you wait for the obvious moment, you often pay the obvious price.

There's no moral here about being heroic or perfectly contrarian. There's a practical point: if you plan to hedge systematically, you should expect to spend much of your time buying protection when it feels slightly ridiculous. That's the cost of getting a decent quote.

How the gap shows up in real P&L

The fear premium doesn't show up as a line item on your brokerage statement labeled "markup." It shows up as a pattern: you buy options, nothing terrible happens, and the options decay or expire. Repeat that enough times and you start to think hedging is a scam.

It isn't a scam. It's insurance behaving like insurance.

If you own a house, you probably pay premiums for years without a major claim. The "loss" is the point: you're renting certainty. In options, the twist is that the premium isn't just paying for expected claims. It's paying for the seller's need to survive the worst-case scenario. In many markets, the average buyer is paying for someone else to warehouse tail risk.

That's why the same hedge can be a hero in a crash and a drag in a normal year. The drag is the premium you paid for a catastrophe that didn't arrive on schedule.

Why implied volatility is not a forecast (and why that's good news)

It's tempting to treat implied volatility like a prediction: "IV is 20%, so the market expects a 20% annualized volatility." That's not quite the right mental model. IV is a *price-implied parameter*, not a pure belief. It blends expectation, supply and demand, hedging pressure, and the risk appetite of the marginal seller.

That sounds like a technical nitpick until you realize what it implies: IV can be "too high" or "too low" relative to what actually happens, and it can stay that way for a long time. It can move even if the underlying price hasn't moved much, because what's changing is the market's willingness to underwrite risk.

For hedgers, this is good news because it means you can sometimes buy insurance in a relatively generous moment. You're not doomed to always pay peak panic prices. But it's also bad news because you can't rely on IV as a clean, stable forecast. It's a mood ring that also doubles as an invoice.

The hidden asymmetry: volatility clusters, and fear jumps

One reason the fear premium exists is that volatility doesn't behave like a polite statistic. It clusters. Calm periods tend to beget calm. Turbulent periods tend to beget turbulence. And when fear arrives, it often arrives in a hurry.

Realized volatility is smooth in the way an autopsy report is smooth: it tells you what happened, neatly. Implied volatility is jumpy because it is forward-looking and defensive. Option sellers don't have the luxury of waiting for realized volatility to confirm danger. They reprice on hints: a credit spread widening, an ugly gap down, a liquidity wobble. Even rumors can widen option markets and lift IV.

That's why implied volatility often spikes before the most dramatic realized volatility shows up in the data. Markets don't wait for your spreadsheet to update.

A simple way to compare them without drowning in math

You don't need to annualize anything or memorize formulas to get value from the implied-vs-realized idea. You just need a consistent habit of comparison.

Pick the time horizon that matches your hedge. If you're buying one-month puts, compare:

- what the option market is charging *now* (implied volatility for that tenor), versus

- what the market has actually delivered over recent one-month windows (realized volatility).

You're not trying to win a forecasting contest. You're trying to answer a business question: *Am I paying a normal insurance premium, or am I paying a panic surcharge?*

Even without a chart, you'll start to feel regimes:

- In complacent regimes, IV can be modest and put protection can feel "reasonable," even though it still bleeds.
- In anxious regimes, IV can rise quickly, and protection can become expensive precisely when everyone wants it.

That regime awareness is the difference between a hedging plan that survives and a hedging plan that gets abandoned at the worst possible moment.

How ``fear premium" interacts with the hedge you choose

The fear premium doesn't tax every protection style equally.

If you buy straightforward puts repeatedly, you are often a direct payer of the volatility risk premium. You're paying the insurer, over and over, for the right to file a claim in a world where claims are rare but brutal.

If you reduce premium outlay by selling something (for example, selling a call in a collar structure), you're partially financing the fear premium with someone else's optimism premium. That trade has its own consequences—you're giving up upside—but it changes the cash flow mechanics of living with hedges.

If you use spreads (buy one put, sell another lower-strike put), you're explicitly choosing to stop insuring the most catastrophic part of the tail in exchange for paying less fear premium. That's not "worse"; it's a definition of coverage. It's the difference between a deductible you can live with and one you can't.

The key is to see the premium as a market price for fear and to recognize that you can alter how much fear you buy. You can't eliminate fear from the bill; you can only decide how much of it you want to carry.

The practical takeaway: treat implied volatility like a price tag, not a prophecy

A long-term investor doesn't need to become a volatility trader to benefit from this. One habit is enough:

Before you buy protection, look at implied volatility for the expiration you plan to use and ask: does this feel like a normal quote or a stressed quote? Then size the hedge as if the quote could get worse right when you need to renew it.

That one question keeps you from building a plan that only "works" when markets are calm. It also keeps you from confusing "expensive" with "bad." Sometimes expensive protection is expensive because risk is genuinely elevated. The point is to recognize when you're paying a fear premium—and to decide consciously whether the emotional relief and portfolio stability are worth the markup today.

3.3 Navigating the Volatility Surface

Option prices don't just change when the market moves. They change *differently* depending on which part of the option chain you're looking at—as if the market has neighborhoods with different insurance rates. A put that protects you 5% below today's price might be merely pricey. A put that protects you 25% below can be downright rude. That's not because the contract is exotic. It's because the crowd that buys it is predictable: when people get scared, they don't reach for "a little protection." They reach for *crash* protection, and they do it all at once.

The map of those different insurance rates is called the *volatility surface*. If that sounds like something you need a quant desk to understand, relax. You don't need 3D plots or graduate math. You just need to know what the surface is trying to tell you in plain English: *the market charges different implied volatilities for different strikes and different expirations*. That's it. Different prices for different flavors of fear.

Think of implied volatility as a menu price, not a single number

Many investors first learn volatility as one number: "IV is 18%," like the market posts a single weather forecast. But open an option chain and you'll see the lie immediately. The implied volatility for a 30-day at-the-money option is often not the same as the implied volatility for a 30-day far out-of-the-money put. Same underlying, same expiration, different strikes—different prices for uncertainty.

The surface is the full menu:

- Across *strike prices*, implied volatility often tilts—that's the *skew* (or *smirk*).
- Across *expirations*, implied volatility can rise or fall—that's the *term structure*.

A single implied volatility number is a snapshot. The surface is the whole landscape.

Once you start thinking in surfaces, you stop asking "Is volatility high?" and start asking the more useful question: *Which volatility is high? The one attached to the protection I actually want?*

Why the left side of the chain is expensive: put skew in one sentence

In equity and index options, downside protection is structurally more expensive because out-of-the-money puts tend to trade at higher implied volatilities than comparable out-of-the-money calls. That pattern is called *put skew*. Some people call it a *volatility smirk* because the curve looks like the market is making a face: "Sure, I'll insure you... for a price."

This is not a tiny detail. It's the difference between buying a hedge that feels like a reasonable bill and buying one that feels like you're funding someone else's retirement.

The market learned to fear the left tail (and never unlearned it)

There's a before-and-after in the way equity options are priced, and the dividing line is the era when markets learned that crashes can arrive faster than models assumed. A severe, sudden drawdown is not just "a big down day." It's a liquidity event, a forced-selling event, a correlation event—a day when everything you thought was diversified starts breathing in unison.

After that lesson, the market stopped treating up and down as mirror images. Upside surprises are fun. Downside surprises can be existential. Option sellers price that asymmetry because they've lived it. Even if you haven't.

The result is a permanent scar in the option chain: deep downside strikes carry an extra premium relative to what a neat symmetric world would suggest. It's as if the insurer remembers every claim anyone has ever filed, even if *you* have never filed one.

Supply, demand, and the fact that ``selling puts" is not emotionally neutral

There's also a simpler explanation that doesn't require history lessons: the natural flow of money is lopsided.

Many institutions are long equities and short fear by default. When they decide to hedge, they tend to buy puts. That creates steady demand for downside insurance.

The natural sellers of that insurance are fewer, and they're often professional volatility sellers, market makers, or structured-product desks that have their own hedging constraints. They don't just sell you a put and go to lunch. They hedge dynamically, they manage inventory, they worry about gap risk, and they require compensation for the times when hedging becomes a contact sport.

Put selling also carries a particular kind of reputational and emotional risk. Selling a call feels like "I'll give up some upside." Selling a put feels like "I'm insuring the building." When the building burns, everyone notices. That social reality shows up in pricing. Markets are made of humans, and humans demand more pay for scary jobs.

Crash protection is expensive because crashes are not just large moves---they are fast moves

A common misunderstanding goes like this: "A 20% rally and a 20% crash are symmetric, so puts and calls should be similarly priced." That would be plausible if markets moved like a sine wave.

Crashes don't arrive like that. They often arrive with speed, gaps, and feedback loops. The difference between drifting down 20% over a year and dropping 20% in a month is everything to an option seller. Fast moves break hedges, widen bid–ask spreads, and force trading in the worst conditions. That *speed* component is part of why out-of-the-money puts carry extra implied volatility.

Even if the eventual magnitude were the same, the path is not. And options care about the path because the probability of ending below a given strike depends on how violently prices can travel.

Reading skew without overthinking it: what it says about insurance pricing

Skew is the market's way of saying: "I'm more worried about bad outcomes than good ones."

When skew is *steep*, low strikes have very high implied volatility. Downside insurance is not just expensive; it is *especially* expensive in the tail. In practical terms, buying far out-of-the-money crash puts is paying a premium for a premium. You are buying coverage where the market's fear is most concentrated.

When skew is *flatter*, the implied volatility differences across strikes are smaller. Downside insurance is still not free, but the "crash surcharge" is less severe. Flatter skew can make certain hedges (especially spreads) more efficient because you're not overpaying as dramatically for the tail.

Skew is not a moral judgment. It's a price. Your job is to decide whether you want to pay it, avoid it, or partially finance it by changing the structure of your hedge.

The term structure: why the same hedge can be cheap in six months and brutal next week

The second dimension of the surface is time. Implied volatility for short expirations can be very different from longer expirations.

In calm markets, longer-dated implied volatility is often higher than short-dated implied volatility. That makes intuitive sense: more time means more uncertainty.

In stressed markets, something interesting happens. Short-dated implied volatility can explode upward. The market is screaming about the next few days or weeks—earnings, a policy decision, a credit event, a geopolitical shock—and the front end becomes the most expensive part of the curve.

This is why "rolling" a hedge can be psychologically tricky. You might have bought a three-month put at a tolerable implied volatility. When you go to renew it a month later, the same maturity window might be priced at a completely different level because the market's short-term anxiety has changed. The surface moved under your feet.

The practical takeaway is not "always buy long-dated options" or "always buy short-dated options." It's simpler: *time is not just a clock; it's a pricing regime.* The market can charge very different rent depending on how close you are to the next potential panic.

Sticky strike vs. sticky delta: two mental models that keep you sane

When the market drops, your strikes don't move, but their meaning changes. A $90 strike put is "far out-of-the-money" when the stock is $120. It's suddenly "near the money" when the stock is $95. That shift matters because implied volatility is quoted by strike, but traders often think in terms of *moneyness* (how far the strike is from the current price) or *delta* (roughly, how sensitive the option is to a move in the underlying).

Practitioners sometimes use two heuristics to describe how the surface "moves" when spot price moves:

- *Sticky strike*: implied volatility at a given strike stays roughly the same even as the underlying moves.
- *Sticky delta* (or sticky moneyness): implied volatility tends to stay tied to options of similar moneyness/delta, meaning the "high IV zone" moves along with the price.

You don't need to pick a side like it's a sports team. The value of these heuristics is psychological: they prevent you from being surprised by something that feels like a pricing glitch.

If the market drops and suddenly your desired put strike looks much more expensive, it may not be "the market ripping you off." It may be that the strike you're looking at has moved into the part of the surface where insurance is always priced higher because it's closer to being needed. The same contract is now a more serious claim on protection.

What skew means for the long-term investor: you can't buy the perfect hedge at a fair price

The dream hedge is simple: pay a small premium, keep all your upside, and be protected against catastrophe. The volatility surface is the market's polite way of saying no.

If you want deep crash protection with puts, the skew means you often pay a lot for strikes that are far below the market. You are buying the most coveted coverage in the chain, and the premium reflects that.

This doesn't mean you should abandon crash protection. It means you should stop expecting it to be cheap in the same way you stop expecting beachfront property to be cheap. The price is not random; it reflects where demand congregates.

Once you accept that, you can make cleaner choices:

- Pay up for the tail because your objective is survival in a very specific nightmare scenario.
- Buy a closer-to-the-money hedge that pays sooner in a moderate drawdown, accepting that it costs more upfront but wastes less time waiting for an apocalypse.
- Use a put spread to reduce exposure to the most expensive part of skew, accepting a cap on protection below the sold strike.

Each of these is a coherent insurance policy. The incoherent policy is the one that assumes the skew won't matter.

A concrete thought experiment: the ``same distance" call vs. put

Imagine a broad equity index at 100. Consider two 3-month options that are the same distance away:

- A 90-strike put (10% down protection).
- A 110-strike call (10% up speculation).

In a symmetric universe, you might expect their implied volatilities to be similar. In the real equity universe, the put often carries higher implied volatility than the call. That means the put is priced as if the market expects fatter tails on the downside—or, more precisely, as if the market is willing to pay more to insure the downside than to gamble on the upside.

Now translate that into your hedging costs. If you buy puts as insurance, you are buying the option type that the market systematically prices rich. If you fund those puts by selling calls, you are often selling the option type that the market prices less rich. The collar can still be a sensible trade, but it's not "free." You're swapping the expensive thing for the less expensive thing. The surface makes that trade visible.

When the surface itself becomes the risk

A hedge can fail in two ways. The obvious way is that the market doesn't fall enough and you lose the premium. The subtler way is that you design a program that requires you to renew protection frequently, and then you discover that renewal happens at the worst possible time.

During stress, the surface can change shape quickly:

- Short-dated implied volatility can spike, making near-term insurance brutally expensive.
- Skew can steepen, making tail strikes even more expensive relative to at-the-money.
- Bid–ask spreads can widen, turning theoretical "fair value" into a practical execution tax.

Even if you *know* hedging is insurance, it can be emotionally destabilizing to watch the insurer raise prices in the middle of the storm. The market does it for the same reason a coastal insurer raises premiums after a hurricane season: the risk is no longer an abstract possibility.

Actionable takeaway: shop the surface the way you'd shop an insurance policy

Before you buy a put hedge, take five minutes and treat the option chain like an insurance quote comparison:

- Look at multiple strikes (closer-to-the-money and further out) and notice how implied volatility changes. That slope is the skew you are paying for.
- Look at multiple expirations (near-term and longer-term) and notice where implied volatility is most inflated. That shape is the term structure you are renting.
- If tail strikes look disproportionately expensive, seriously consider a put spread as a way to avoid overpaying for the most fear-loaded part of the chain.

The goal isn't to find a "cheap" hedge. The goal is to buy coverage where the surface isn't quietly charging you a luxury tax for the exact strikes everyone else panics into.

3.4 Essential Greeks for Hedgers

Buying protection with options is like buying a vehicle with an unfamiliar dashboard. The car moves, but the gauges don't mean what you think they mean. You press the pedal (buy a put) and the speed changes (your portfolio risk shifts), but then the road tilts (volatility jumps), the engine revs (your hedge ratio morphs), and fuel mysteriously disappears even when you're parked (time decay). The *Greeks* are simply the labels on those gauges. Not equations to worship—signals to pay attention to before you end up stranded on the shoulder, arguing with your own trade.

What matters for hedging is not a full Greek alphabet. A long-term investor can do serious risk management with four: Delta, Gamma, Vega, and Theta. They answer the questions you actually care about when markets get unfriendly:

- *How much protection do I have right now?* (Delta)
- *How quickly does that protection change as prices move?* (Gamma)
- *How much do I benefit when fear spikes?* (Vega)
- *What am I paying each day to hold this?* (Theta)

If you can read those four gauges, you can size hedges with intention instead of superstition.

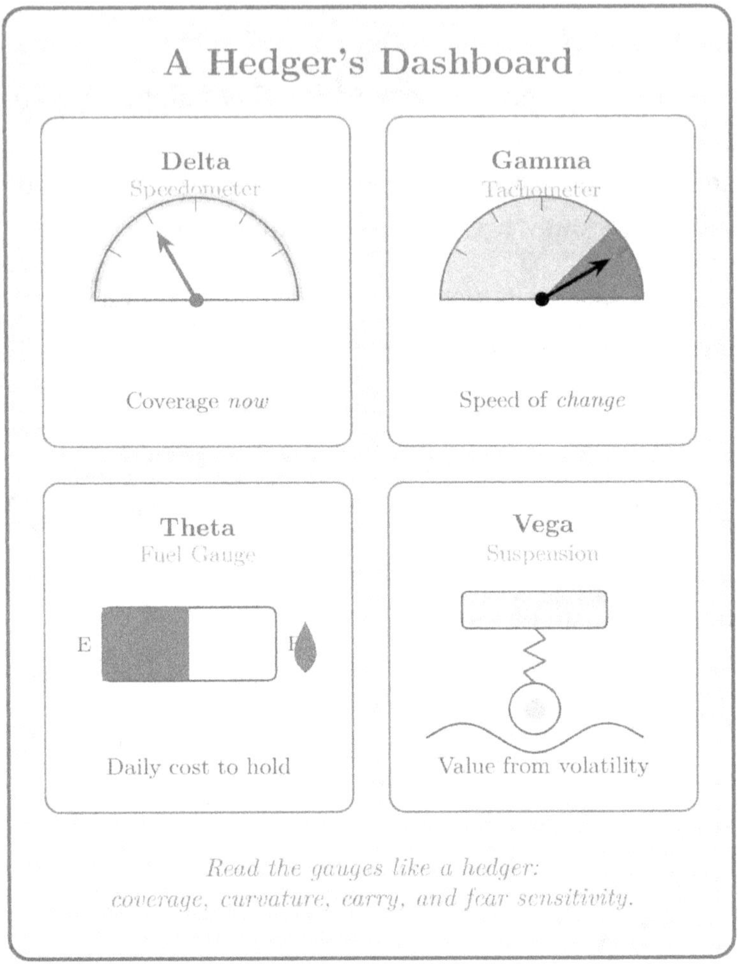

Delta is the speedometer because it tells you, right now, how much your option position moves when the underlying moves. If you own a stock, its delta is effectively 1: a $1 move in the stock is a $1 move in your position per share. Options aren't so literal. A put's delta is negative because it gains value when the underlying falls. When you buy puts to hedge, delta is the first number that answers the question you actually mean when you say "am I protected?": *If the market drops today, how much does my hedge respond today?*

A delta of −0.20 on a put (roughly) means the option might gain about $0.20 for a $1 drop in the underlying, *right now,* all else equal. In portfolio terms, delta is your immediate shock absorber. It's also why many "cheap" far out-of-the-money puts disappoint people in normal drawdowns: their delta can be tiny until the market gets close enough for the insurance to become relevant.

The trick with delta is that it can seduce you into false precision. Delta is not a promise. It's a local estimate. Markets don't move in polite $1 increments, and "all else equal" is the least reliable phrase in finance. In a selloff, implied volatility tends to rise and the surface can reshape, so the hedge can behave better (or sometimes stranger) than a simple delta would suggest. Still, delta is the cleanest way to translate option jargon into a sentence a human can use: *this is how much protection I have at this price level.*

Gamma is the tachometer because it tells you how quickly delta itself changes when the market moves. Delta is speed; gamma is acceleration. If delta answers "how much protection do I have right now," gamma answers "how quickly does that protection *increase* if things get worse?" For a hedger, that's the difference between a seatbelt and an airbag.

High gamma means your hedge becomes more responsive as the market moves toward the strike. That's exactly what makes options attractive as protection: they can become more protective at the moment you most need them. But gamma is not a free lunch. High gamma often comes packaged with higher theta (more on that in a moment). The market charges you for convexity—for the privilege of owning a position that curves in your favor.

Gamma also explains why option hedges can feel calm and then suddenly feel alive. Many investors watch a hedge "do nothing" for weeks and conclude it's broken. Then the market drops and the hedge starts moving faster than expected. That's gamma showing up. The option isn't being dramatic; it's doing what it was built to do: change character as the underlying approaches the danger zone.

There's a practical way to use gamma without calculating anything complicated: notice where your strikes sit relative to the market. If your protective put is far below current prices, its gamma is often low at today's level—meaning it won't "wake up" quickly unless the market falls a lot. If your hedge is near the money, gamma is higher and the hedge will respond more sharply to moves. This is why "distance to the strike" is not just about payoff diagrams; it's about how quickly your insurance transitions from theoretical comfort to actual protection.

Vega is the suspension sensitivity because it tells you how much your option's price changes when implied volatility changes. In plain language: vega measures how much you benefit when fear gets louder. If implied volatility rises, long options (like long puts) tend to gain value. That's the part of hedging that can feel almost magical when it works: the market drops, everyone panics, implied volatility spikes, and your hedge gains value not only because the underlying fell but because the insurance itself got more expensive.

But vega is also where many hedgers get surprised in the opposite direction. If you buy protection when implied volatility is already high, and then the market stabilizes, implied volatility can fall even if prices don't rebound much. Your put can lose value because fear drained out of the system. You'll hear traders call this "vol crush," and it can feel unfair: you bought insurance, the danger didn't disappear, but the premium still shrank because the crowd exhaled.

Vega is also a reminder that not all "volatility changes" are equal. Implied volatility does not move as one uniform blob. Different expirations move differently, and different strikes can reprice in different ways when markets gap or grind. That's why vega is best treated as a sensitivity to the *market's pricing of uncertainty*, not a sensitivity to your personal sense of dread. The market can become less afraid even while you remain uneasy, and your hedge will mark to *their* mood, not yours.

Theta is the fuel gauge draining because it measures time decay: how much value the option loses with the passage of time, holding other factors constant. Theta is the most honest Greek because it is relentlessly, almost insultingly, predictable. If you are long options for protection, theta is usually working against you. It is the daily rental cost of owning convexity.

Theta is also why hedging can feel like a slow leak in performance during calm markets. Nothing is "wrong." You are paying for coverage and not filing a claim. The premium bleeds out over time, and the bleed accelerates as expiration approaches. If you've ever watched a short-dated put lose value in a sleepy market and felt personally mocked by your brokerage app, that was theta, doing its job.

The most useful way to think about theta is not as a punishment but as a budgeting tool. Every hedge has a carry cost, and theta is the part you can't negotiate away. You can choose longer expirations to slow the drain, but you pay more upfront. You can choose shorter expirations to reduce upfront premium, but then you accept faster decay and more frequent re-buying. Theta forces you to be honest about how long you can afford to rent protection.

The Greeks are not independent gauges; they're a tightly coupled machine. One of the most important couplings for hedgers is the gamma—theta tradeoff. High gamma often means more theta bleed. The market charges a higher daily rent for a position that responds aggressively to moves. That's why "I want a hedge that really kicks in during a selloff" often translates, in practice, to "I am willing to pay more carry for higher convexity." If you aren't willing to pay the carry, you often end up with protection that only helps in very extreme scenarios.

Vega and theta have a relationship too, and it's the relationship that makes hedging feel emotionally confusing. You can be *right* about risk and still lose money on your hedge if the market becomes less afraid while time passes. A hedge is not just a bet on direction. It's a bundle: direction (delta), curvature (gamma), fear sensitivity (vega), and rent (theta). You can win one of those and lose another in the same week.

A simple real-world picture: suppose you buy puts because you suspect a rough quarter ahead. If the market drifts down slowly, your puts may gain from delta but lose from theta, and implied volatility might not rise much, so vega doesn't help. If the market drops sharply, gamma and vega can team up with delta and you can get the kind of payoff that makes hedging feel brilliant. Same thesis ("downside risk"), very different path, very different Greek mix.

Greeks are also the antidote to a common sizing mistake: buying a hedge based only on premium budget. "I can afford $2,000 of puts per quarter" is a real constraint, but it's not a hedge ratio. Delta helps convert that budget into *coverage*. Two hedges that cost the same can deliver radically different delta exposure depending on strike, time, and volatility. The investor who watches only premium is shopping insurance by monthly bill, not by deductible and coverage limits. The Greeks are how you read the policy.

For long-term investors, one especially practical move is to treat delta as the headline number and use the other Greeks as the fine print. Start by deciding what fraction of your equity exposure you want to dampen in an immediate selloff. That's a delta question. Then check gamma to see whether your hedge becomes meaningfully stronger if the market continues lower (useful for drawdown control). Check vega to understand whether you're relying on a volatility spike to make the hedge work (and whether you're buying that vega at an expensive moment). Finally, check theta to see the daily carrying cost of keeping the protection alive.

One more reality check keeps Greek-watching from becoming an obsession: every Greek is model-dependent to some degree, and the market can move in discrete jumps that don't respect smooth sensitivities. Treat the Greeks as a dashboard, not a GPS. A dashboard helps you avoid doing something obviously reckless. It does not guarantee you arrive on time.

Actionable takeaway: run a 60-second ``dashboard check" before you place the trade

Before you buy (or roll) a protective option position, pause and do a quick, repeatable check:

- *Delta:* If the market drops 1% today, does this hedge noticeably respond, or is it mostly a lottery ticket?
- *Gamma:* If the drop continues, does my protection *accelerate*, or does it stay sluggish until a catastrophe?
- *Vega:* Am I paying a premium for fear that's already expensive, or am I buying volatility sensitivity while it's still reasonably priced?
- *Theta:* Can I afford the daily bleed for as long as I plan to hold this hedge, or will I be forced to abandon it out of annoyance?

If any answer makes you uncomfortable, change the strike, change the expiration, or change the structure—before the market changes it for you.

3.5 Calculating the All-In Cost

A hedge is one of the few things in personal finance that can be simultaneously *expensive*, *useful*, and *correct*—and still make you feel foolish. The reason is simple: the bill you see is not the bill you pay. The quoted premium is the sticker price. The *all-in cost* is the sticker price plus the slow drips and hidden tolls that show up only after you've lived with the hedge for a while.

Insurance works this way outside markets, too. Your car insurance premium is only the beginning: deductibles, exclusions, and rate increases after claims matter. Options just compress that reality into a cleaner, harsher form. You pay in cash up front, and then you pay again in time decay, trading friction, and the opportunity cost of capital that could have been compounding elsewhere.

The good news is that none of this requires a PhD. You can estimate the real cost of a hedging program with a few numbers you can pull from an option chain and your own portfolio. The goal isn't to compute a perfect answer. The goal is to stop being surprised.

Start with the only cost you can't argue with: premium out the door

The simplest component of hedge cost is the premium you pay to buy options. If you buy one put contract at $4.00, and the contract multiplier is 100, you paid $400. That $400 is not "at risk." It is *spent*. It can come back to you as profit if the hedge pays off, but as a budgeting reality it's already gone.

This framing matters because investors often treat premium like an investment they are "down on" when it decays. Mentally, that encourages the worst behavior: holding a hedge too long hoping it "comes back," or refusing to roll because you don't want to "lock in a loss." A hedge isn't a bruised ego. It's a policy. The premium is the policy cost.

When you build a hedging plan, you're not deciding whether a single put is worth $400. You're deciding whether *repeatedly paying* $400 (or whatever the number is) is worth the reduction in drawdown risk you're buying.

Theta bleed: the rent that becomes visible only after you sign the lease

If premium is the sticker price, *theta* is the monthly payment. For long-option hedges, theta is usually negative: your option loses value each day, all else equal. Investors feel this as "bleed," and it's the most consistent source of drag in a hedging program.

The key insight is that bleed is not evenly spread through time. As expiration approaches, theta tends to accelerate. Short-dated protection can feel like it melts. That doesn't automatically mean short-dated hedging is wrong; it means the cash-flow profile is demanding. You are signing up for a faster rent schedule and more frequent renewal decisions.

A practical way to price theta in human terms is to convert it into an annualized budget. If your hedge bleeds, say, $8 per day across all contracts, that's roughly $2,000 per year. Suddenly the hedge stops being an abstract "options thing" and starts being what it really is: a line item competing with savings, travel, and retirement contributions.

Even better: theta bleed can be estimated *before* you place the trade because most brokerage platforms show theta per contract. You don't need to predict the future to see the rent. You only need to decide whether you can afford it.

Roll cost: the recurring ``renewal tax" that makes good hedges look bad

A hedge you intend to keep must be renewed. Options expire. That expiration is not just a date; it's a forcing function. It pushes you into the market to buy new protection whether you feel like it or not.

That renewal is where costs compound:

- You pay premium again.
- You often pay spreads again.
- You may be forced to buy at elevated implied volatility if markets are stressed.

The painful irony is that your hedge is most valuable when markets are dislocated—and that's also when it can be most expensive to replace. In a panic, implied volatility spikes and skew steepens. The market raises the price of insurance while you're actively trying to renew your policy. This is not an inefficiency; it's the options market functioning like an insurance market. After storms, premiums go up.

So the all-in cost of hedging is not "what did I pay for the put?" It's "what is the expected cost of repeatedly buying protection across different volatility regimes, including the bad regimes when I'm least excited to pay?"

Bid--ask spreads: the cost of crossing the street in bad weather

The bid–ask spread is the difference between what you can sell at (bid) and what you can buy at (ask). It's easy to ignore when markets are liquid and spreads are tight. It becomes impossible to ignore when markets are fast, dealers widen quotes, and liquidity evaporates.

For hedgers, spreads matter in three ways:

- You often pay the ask when you buy protection.
- You often receive the bid when you sell or roll it.
- In stressed periods, spreads widen, so the same "paper value" becomes harder to realize.

A simple estimation method keeps this grounded. Take the quoted spread per option (ask minus bid), multiply by 100 (contract multiplier), and then multiply by the number of contracts you expect to trade per year (including rolls). If the spread is $0.15, that's $15 per contract per trade. Two trades per roll (sell old, buy new) is $30 per contract per roll, before commissions and fees. Over time, those small numbers become a meaningful drag.

Spreads are also where "cheap options" can be expensive in disguise. Deep out-of-the-money puts might have low premiums, but if they trade with wide spreads, you're paying a hefty friction cost relative to the premium. It's like buying a low-cost appliance that can only be serviced by one shop across town.

Slippage and execution: the real price is what you get, not what you saw

Even if the spread looks tolerable, you can still pay more than expected through *slippage*: getting filled at a worse price than you intended because the market moved or liquidity was thin.

This is especially relevant for protective hedges because many people place orders when they are emotionally activated—during a drop, after a scary headline, when volatility is surging. That's the worst time to act like a tourist in the options market. Market orders can be punishing. Limit orders can miss. The cost of impatience is real.

The "all-in" mindset suggests a sober practice: if hedging is part of your risk plan, treat execution as part of the hedge, not as an afterthought. The most beautifully designed strategy can be made mediocre by consistently sloppy fills.

Opportunity cost: the stealthiest expense because it never shows up on a statement

Premium, theta, spreads—those show up as losses you can point to. Opportunity cost is sneakier. It's the return you didn't earn because money was tied up paying for hedges or sitting in cash to make hedging feel comfortable.

Even when options are paid fully in premium (no margin), the capital used for premiums is capital that cannot compound in your core portfolio. Over long horizons, that matters. A steady 1%–2% annual drag is not "small." It's a compounding leak.

There's also a behavioral opportunity cost: some investors hedge in a way that makes them feel so safe that they increase equity exposure or take more concentrated bets elsewhere. The hedge becomes an emotional permission slip. Sometimes that's intentional and rational. Sometimes it turns the hedge into a fragile talisman: "I can do anything because I bought puts." That is how opportunity cost can come back as actual cost.

A practical way to incorporate opportunity cost without pretending you can forecast markets is to set a *hedge budget* as a percentage of portfolio value per year and treat it like an insurance premium you're willing to pay for smoother long-term outcomes. If your budget is 1% annually, then you need a hedging approach whose expected premium plus friction plus roll costs roughly fit inside that envelope over time.

The ``cost of carry" for a hedge: a simple, usable formula

You can think of hedging cost the way professionals think about carry: what it costs to maintain a position through time.

For a long-put hedging program, an investor-friendly approximation looks like this:

$$\begin{aligned}\textit{All-in annual hedge cost} \approx\ & \textit{Premium spent} + \textit{Realized roll losses} \\ & + \textit{Spreads/fees} + \textit{Opportunity cost}\end{aligned}$$

That is intentionally not precise. It's a checklist that prevents self-deception.

If you want a slightly more operational way to calculate it, do it cycle by cycle. Suppose you roll quarterly:

1. Record total premium paid for the hedge at initiation.
2. At the roll date, record what you receive when you close the hedge (or let it expire).
3. The difference (premium paid minus premium recovered) is your net option cost for that cycle.
4. Add estimated execution friction: half-spread on entry + half-spread on exit (or your actual fills).

5. Repeat across cycles and annualize.

Notice what's not required here: predicting volatility, forecasting crashes, or arguing with yourself about whether you "should have" bought at a different time. This is accounting, not hindsight.

Once you have a year of these numbers, the strategy stops being a philosophy and becomes a cost profile you can evaluate like any other expense.

Why ``cheap protection" can be the most expensive kind

Many investors learn the hard way that cheap protection can be expensive for two opposite reasons.

First, very cheap puts are often far out-of-the-money. They can bleed for months and then expire worthless through multiple drawdowns that never quite reach the strike. You pay repeatedly and never get the emotional relief you expected because the hedge rarely pays in the scenarios you actually experience.

Second, very cheap options can be illiquid. The spreads can be wide, the fills can be poor, and the position can be hard to adjust when you most need flexibility. The savings in premium are then quietly handed back in execution costs.

"Cheap" is not a feature. *Efficient* is the feature. Efficient protection is protection that delivers meaningful drawdown relief for an all-in cost you can sustain through boring years and scary years alike.

A real-world reality check: exercise, assignment, and settlement can affect the bill

Options are standardized contracts with specific rules about exercise, assignment, and settlement. Those mechanics don't usually change the day-to-day mark-to-market, but they absolutely can affect real-world outcomes when positions get deep in-the-money, close to expiration, or paired with other positions.

For many hedgers, the cleanest way to avoid unpleasant surprises is to treat options like perishable goods: manage them before expiration, don't leave critical decisions to the last trading day, and understand whether your contract is equity-settled or cash-settled (especially for index options). If you're using options as insurance, you want the claim process to be predictable.

This isn't about fear-mongering. It's about professionalism. A hedge is supposed to reduce stress, not create a new category of it.

Actionable takeaway: run a ``hedge cost audit'' before committing to a program

Before you commit to any ongoing hedging approach, do a short audit and write the answers down:

- *Budget:* What percentage of portfolio value per year are you genuinely willing to spend for protection without abandoning the plan?

- *Bleed:* Based on theta and your intended holding period, what's the expected decay cost per month in calm markets?
- *Roll plan:* How often will you renew, and what happens if implied volatility is elevated when you must roll?
- *Friction:* What are the typical bid–ask spreads and how many contracts will you trade per year?
- *Opportunity cost:* If the hedge costs 1%–2% per year, what are you giving up, and is the smoother ride worth it?

If those numbers don't fit together on paper, they won't fit together in real life. The goal isn't to find a hedge that pays for itself. The goal is to buy survival at a price you can keep paying when the market is calm enough to make you question why you ever wanted insurance in the first place.

Chapter 4

Core Hedging Structures

If you wouldn't drive a car without insurance, why drive your retirement savings into a volatile market unprotected? In this chapter, we open the toolkit. We aren't just memorizing definitions; we are shopping for the right policy. Some hedges are like a concrete bunker (expensive but bombproof), while others are like a sturdy umbrella (good for rain, bad for hurricanes). We will explore how to construct these safety nets, how much they cost, and how to choose the one that matches your specific 'sleep-at-night' number.

4.1 Defining Hedging Objectives

The most expensive hedge is the one you didn't need.

It's easy to picture a hedge as a product you *buy*: a put, a collar, an ETF with "buffer" in the name. But a hedge is really a sentence you write about your future self. It answers one question: *What kind of pain are you refusing to tolerate?* Not "How much downside do I want to reduce?"—that's vague enough to justify almost anything. The useful version is specific: *"I don't want to be forced to sell in a 2008-style drawdown."* Or: *"I can handle losses, but I can't handle the sleeplessness of a portfolio that whipsaws 3% a day."* Or: *"I'm fine with normal volatility; I just don't want a drawdown deeper than 15% while I'm paying a tuition bill."*

Hedging without an objective is like buying insurance without naming the disaster. You'll either overpay for coverage you never use, or you'll discover—right when you need it—that you insured the wrong thing. Even worse, you can accidentally hedge the part of investing that was doing the heavy lifting. Many "protective" products don't protect your long-term plan; they protect your short-term feelings by quietly selling away your upside. That can be a perfectly rational trade—*if* you meant to do it.

A practical way to set objectives is to sort your fear into one of three buckets. Not because life fits neatly into buckets, but because each bucket pushes you toward a different hedging structure, a different budget, and a different definition of success:

- *Catastrophic crash protection*: "If it happens, I want a hard floor."
- *Volatility smoothing*: "I want the ride to be less violent, even if I give up something for it."
- *Shallow drawdown limits*: "I'm not trying to survive the apocalypse; I'm trying to prevent a normal correction from becoming a personal financial event."

The buckets sound similar until you attach dates, dollar amounts, and a realistic description of your behavior under stress. The portfolio doesn't care about your intention; the broker doesn't care about your intention; the market certainly doesn't. Only your objective turns a hedge from a random set of option contracts into a design.

Objective #1: Catastrophic crash protection (``I want a floor")

Catastrophic protection is the financial equivalent of not wanting to discover, mid-flight, that your parachute has a deductible.

This objective usually isn't about maximizing returns. It's about avoiding a forced, irreversible decision: selling near the bottom because you *must* raise cash; liquidating because your margin loan is calling; abandoning a long-term allocation because a drawdown is psychologically intolerable; or simply losing the ability to take advantage of bargains because you're defending the rent. A true crash hedge is less about being right and more about staying intact.

Here's a clean way to describe the objective in one line:

> *Over a defined window (say, the next 6–18 months), I want to cap my worst-case loss at approximately X% on the portion of my portfolio that matters.*

That "portion that matters" phrase is not a loophole—it's the point. Your portfolio is not one lump. It's a set of promises. The money you need to keep your lifestyle stable over the next year is not the same money you plan to leave invested for twenty. Catastrophic protection usually belongs on the part that would trigger life-changing decisions if it were cut in half.

A common mistake is to describe a crash objective as "I want to be protected if the market falls 50%," and then buy something that only helps if the market falls *exactly* 50%. Real crashes don't land on clean numbers. They gap. They bounce violently. They create "down 6%, up 5%, down 7%" weeks that break people's discipline long before the final low. Crash protection, if it's doing its job, has to be robust to messy paths, not just tidy endpoints.

Two practical tests tell you whether you're truly aiming for catastrophic protection:

Test 1: Would you be satisfied if the hedge lost money for years? A real crash hedge is like a fire extinguisher. You don't brag that it "returned 11% annualized." You accept that most years it looks like wasted money. If you'll only buy a hedge that "pays for itself," you're probably not buying catastrophic protection. You're buying a strategy that *hopes* to be a hedge.

Test 2: Are you trying to protect against a specific calendar window? Catastrophic hedging is often time-boxed: the years around retirement, a business sale, a home purchase, a period of concentrated stock exposure, or simply the era when your human capital is fragile. If there is no window, you may drift into paying permanent insurance premiums on an asset (equities) whose long-term expected return is partly compensation for bearing risk in the first place.

The emotional truth behind this bucket is worth stating plainly: catastrophic hedging is often an antidote to panic. It doesn't just change your payoff diagram; it changes the probability that you'll stick with your plan when headlines turn feral.

Objective #2: Volatility smoothing (``I want a calmer ride")

Some investors don't actually need a hard floor. They need a portfolio that doesn't feel like a hostage situation.

Volatility smoothing is an objective that sounds soft until you live through a year where the market turns into a daily referendum on your intelligence. A portfolio can be up on the year and still be psychologically exhausting if the path is jagged enough. And jagged paths have a nasty side effect: they invite bad timing decisions. People don't usually blow themselves up by holding a diversified portfolio through a crash. They blow themselves up by *changing* the portfolio at the worst possible moment because the ride becomes unbearable.

Volatility smoothing is not the same as "reduce risk." It's closer to "reduce the chance I do something stupid." That makes it a behavioral hedge as much as a mathematical one.

If this is your bucket, define success with a measurement that respects the lived experience:

- smaller peak-to-trough swings over months, not minutes;
- fewer portfolio days that feel like emergencies;
- a return stream that stays investable—one you can keep adding to, rebalancing into, and not abandoning.

There's a hidden trap here: smoothing strategies often accomplish calm by quietly auctioning away your best days. That can be a fair bargain—many people would gladly trade some upside for a portfolio that behaves—but it must be chosen consciously. The market's long-term return is lumpy. A surprising amount of equity performance arrives in bursts. If your smoothing method systematically sells off those bursts, you may end up with a portfolio that feels wonderful right up until you realize it no longer funds your goals.

So the objective must include a line you won't cross on upside. Not "I want upside," but "I am willing to cap my upside beyond roughly X% in a strong year in exchange for a steadier ride." Without that sentence, you'll experience the most demoralizing outcome in hedging: paying for comfort, then resenting the cost the first time the market rips higher without you.

Volatility smoothing is also where misunderstandings flourish because the word "hedge" gets used loosely. Some strategies reduce volatility by harvesting option premiums; others reduce volatility by owning convex protection that kicks in during shocks; others reduce volatility simply by holding more cash or high-quality bonds. They can all be called "hedges" in casual conversation. Your objective is what forces you to pick the right mechanism rather than the right marketing label.

Objective #3: Shallow drawdown limits (``Just don't let a normal hit become a life event")

This bucket is for the investor who can handle equity risk—they just can't afford for timing to be cruel.

Imagine you're doing everything "right." You're diversified. You're long-term. You rebalance. And then the market decides to schedule a 20% decline during the exact year you plan to make a down payment, pay tuition, or step away from full-time work. In theory, a long-term investor should shrug. In real life, that's when the spreadsheet stops being a spreadsheet and becomes a family meeting.

Shallow drawdown hedging is about the first leg down, not the last. You're not trying to win the end-of-the-world lottery. You're trying to cover the kind of drop that happens often enough to be annoying and costly: corrections, fast downdrafts, grinding bear markets that feel "not that bad" until you realize they've chewed through 18 months of progress.

The objective statement here can be delightfully concrete:

> *Over the next year, I want to reduce the impact of a 10–20% market decline on my portfolio by about half, while keeping most of my upside.*

This goal tends to pair with hedges that have a *policy limit*. That's not a bug; it's the whole point. You're saying: "I want help in the neighborhood where bad luck tends to happen, and I'm willing to self-insure the truly extreme tail because it's too expensive to insure fully every year." It's the deductible model in human terms: you can pay the first part of the damage, you just don't want a routine accident to total the car.

If you're honest, this bucket is often about preventing a change in life trajectory. A 15% drawdown in a brokerage account may be emotionally unpleasant but financially survivable. A 15% drawdown in the year you planned to retire can turn into "work three more years." Shallow drawdown hedging is a way of taking the calendar's power away from the market.

The unglamorous constraint: what kind of failure is unacceptable?

After you pick a bucket, add one more sentence that feels boring but saves you from ugly surprises:

> *I cannot tolerate being forced to sell or buy the underlying asset at an inconvenient time.*

That sentence is about *mechanics,* not markets. Options are contracts with rules: expirations, exercises, assignments, settlement conventions, and in some cases early-assignment risk. The market can behave; your hedge can still misbehave operationally.

If you're using hedges built from options you *buy,* the main mechanical risk is expiration. Your protection is time-limited. If you forget to renew it, you're uninsured again—often right when volatility calms down and the insurance feels "obviously unnecessary." That's exactly when people let it lapse.

If you're using hedges that involve options you *sell* (to finance the hedge or generate income), your mechanical risks expand. Short options can be assigned. Corporate actions and dividends can make assignment more likely at precisely the wrong moment. The result can be forced selling (getting called away) or forced buying (being assigned on a short put) that disrupts your plan. None of these are reasons to avoid the strategies; they are reasons to explicitly incorporate operational tolerance into the objective.

A useful way to write this into your objective is to specify not just the outcome you want, but the *constraints* you must respect:

- "I can manage monthly rolls, but I cannot watch options daily."
- "I'm fine being called away; I'm not fine triggering taxable gains this year."

- "I can tolerate a hedge that sometimes over-hedges; I cannot tolerate margin calls."

Constraints are objectives wearing practical clothes.

Budgeting: the part where fantasy meets arithmetic

Every investor wants a hedge that is cheap, powerful, and doesn't touch upside. The market is not a charity. It prices those desires.

So your objective needs a premium budget, stated as a percentage of the portfolio (or of the portion being hedged) *per year*. This single number prevents self-deception. Without it, you can always justify "just a little more protection," until you realize you've built a beautiful fortress and forgotten to fund retirement.

How to pick the budget without pretending you can optimize it perfectly:

- Decide what you're really buying: survival, sleep, or calendar protection.
- Pick a number that is emotionally tolerable to pay in a calm market *and* emotionally tolerable to watch "go to zero" in a year with no crash.
- Treat the budget as a recurring expense, like insurance, not a one-time clever trade.

If that sounds unromantic, good. The romantic version is the investor who discovers, after years of dabbling, that their "hedge" was just a slow leak in performance. The goal is not to be clever; it's to be intentional.

A simple objective worksheet (write it down like a contract)

Here's a template that turns vague anxiety into something you can actually hedge. Write it in plain English—not because the market needs it, but because *you* will need it when the market gets loud.

> *Time window:* I care most about the next __________ months/years.
> *Bucket:* My main goal is (catastrophic floor / volatility smoothing / shallow drawdown limit).
> *Worst-case tolerance:* Over that window, I want my loss on the hedged portion to be no worse than about __________%.
> *Upside trade-off:* I am willing to cap/drag my upside by about __________% in exchange for protection.
> *Budget:* I am willing to spend about __________% per year (or __________ dollars) on hedging costs.
> *Operational constraints:* I cannot tolerate ______________________________.
> *Behavioral promise:* If markets fall fast, my pre-committed action is ______________________________.

That last line—the behavioral promise—is where hedging quietly becomes powerful. A hedge is not only a payout profile; it's a plan for how you will behave under stress. If you don't specify behavior, you may use the hedge as an excuse to trade more, not less.

When objectives clash inside one portfolio

Many portfolios contain competing objectives without realizing it. One part of the account is "forever money." Another part is "money for the next three years." Another part is "the position I'll never sell because it's my employer stock and I'm emotionally attached." These objectives pull hedging in opposite directions.

The fix is not to find the one perfect hedge. The fix is to admit the portfolio is really multiple sub-portfolios living in one statement. You can hedge only what needs hedging. You can leave long-horizon money largely unhedged and protect the money that would cause life disruption if it took a hit. You can also hedge concentrations differently from diversified holdings. This is not complexity for its own sake; it's a way to avoid paying insurance premiums on the wrong dollars.

If you do nothing else, at least separate "I fear losing money" from "I fear *selling* at a bad time." Those are different fears. The market can recover from a drawdown; your plan cannot recover from a forced liquidation at the bottom. Many hedges are most valuable precisely because they reduce the probability of becoming a forced seller.

Actionable takeaway

Before you look at any hedge quotes, write a one-paragraph hedging objective that includes (1) a time window, (2) one of the three buckets—catastrophic floor, volatility smoothing, or shallow drawdown limit—(3) a maximum acceptable loss, (4) an explicit upside trade-off you can live with, and (5) a hard annual budget. If you can't write that paragraph, you're not ready to buy a hedge; you're just shopping for a feeling.

4.2 The Protective Put: Buying the Floor

The protective put is what people *think* they're buying when they say, "I just want to know the worst-case outcome." It's the closest thing markets offer to turning an anxious, open-ended risk ("how bad can this get?") into a number you can write on a sticky note. You pay a premium up front, and in exchange you buy the right to sell at a predetermined price. That right is the floor.

It also has a personality: stubborn, reliable, and unapologetically expensive. If you want crash insurance that doesn't depend on a stop-loss being triggered, a broker being reachable, or liquidity being kind during a panic, you're describing a put option.

The idea, without the jargon

A *put option* is a contract that gives you the right (not the obligation) to sell something at a fixed price (the *strike price*) on or before a specific date (the *expiration*). If the market price falls below the strike, that right becomes valuable. If the market stays above the strike, the put expires worthless—and the premium you paid is the cost of having slept better.

A protective put is simply: *own the asset, and buy a put on the same asset*. You'll also hear "married put," which is a poetic way of saying the put is attached to the position like a seatbelt:

it doesn't matter whether you *feel* like using it later; it's already there.

If you own 100 shares of a stock (or an ETF), one standard option contract typically corresponds to those same 100 shares. That matching of share-for-share is what makes the floor feel so clean. When the market falls, the stock loses value, but the put gains value. Past a certain point, the put's gains begin to offset the stock's losses, and the combined position stops bleeding in the same way.

A concrete story: the moment you stop checking your phone

Suppose you own an ETF at $100 per share. You like it. It's a core holding. You're not trying to trade it. You're trying to *keep* it—but you've also lived long enough to know markets sometimes do that thing where they open down 4% and the news anchor's voice drops an octave.

So you buy a put with a $95 strike that expires in, say, three months. You pay a premium; maybe it's $2 per share. That premium is the admission ticket.

Now picture the week from hell. The ETF drops to $70.

Unhedged, your position is a straight line down: you're down $30 per share. With the put, you have the right to sell at $95. That right is suddenly worth about $25 per share (roughly $95 minus $70), ignoring time value for the moment. Your stock is down $30, your put is up about $25, net down $5—plus the $2 premium you paid. The market can keep falling; the math doesn't start panicking. The put keeps offsetting.

That is the protective put's superpower: it takes the *shape* of your risk and snaps it into a different shape. You go from "I can lose almost everything" to "I can lose roughly this much over this window." It doesn't eliminate loss. It eliminates the bottomless pit.

The floor is real, but it isn't free

There's a reason the protective put is called "portfolio insurance" without anyone laughing. It behaves like insurance in two very important ways:

- It costs money up front, often in a year where you won't "need it."
- The moment it feels obviously unnecessary is often the moment it's cheapest—and the moment you'll be tempted to cancel it.

The premium is not a nuisance; it's the deal. If you want certainty, you rent it. And unlike a one-time home insurance bill, market insurance has a habit of renewing itself in your budget: when the option expires, the protection ends. If you want to stay insured, you buy a new put, a process traders call *rolling*.

This is where many investors get the first honest taste of what hedging really is: not a clever trick, but an ongoing expense that competes directly with compounding. If you spend 2% of the portfolio per year on protection, the long-run cost is not "2%." It's the growth you didn't get to keep. That trade can still be worth it. But it should be made with eyes open.

Meet the three knobs: strike, expiration, and size

Protective puts feel simple, but the moment you try to place a trade you face three knobs that determine everything you care about: cost, protection, and how long the protection lasts.

Strike price (where the floor sits).
A higher strike (closer to today's price) means a higher floor and a higher premium. A lower strike means cheaper insurance but more pain before the protection really kicks in. People often buy protection too far out-of-the-money because it looks "affordable," then feel betrayed when the market falls 12% and the hedge barely helps. That isn't betrayal; it's what you paid for.

Expiration (how long the floor exists).
Short-dated puts can be cheaper in absolute dollars but are brutally perishable. Long-dated puts cost more, but the insurance doesn't evaporate as quickly. There's also a psychological component: short expirations create a recurring decision point where you must renew. In calm markets, renewal feels like buying umbrellas in the desert.

Hedge size (how much of the position you insure).
Insuring 100% of a position is emotionally satisfying and financially costly. Insuring 30% can be a smart compromise: you won't be fully "saved," but you might prevent a drawdown from becoming a behavior problem. Many investors don't need a hedge that makes them whole; they need a hedge that keeps them rational.

The married put, defined precisely

A *married put* is a long stock (or ETF) position combined with a long put option on the same underlying, typically with a contract size that matches the shares owned. The payoff at expiration becomes "stock value, but not below the strike," minus the premium you paid.

Immediately after you internalize that, you should see why it's called buying the floor: regardless of how far the asset falls, you can still sell at the strike (during the life of the option). The market can be dramatic; your minimum exit price is not.

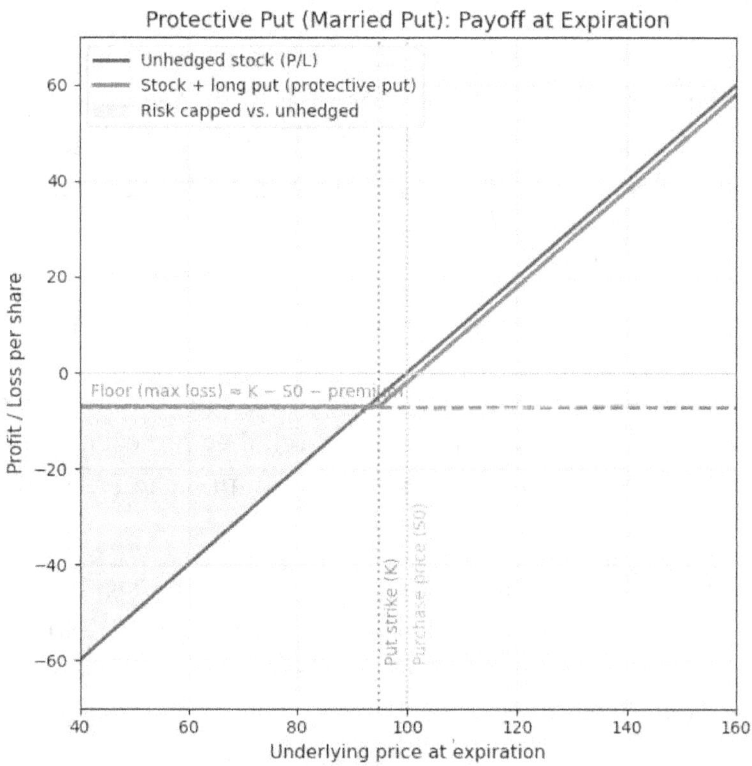

Why a put hedge feels different from a stop-loss

A stop-loss order is a promise you make to yourself: "If the price hits X, I will sell." A protective put is a promise the market makes to you: "If the price falls below X, you still have the right to sell at X (until expiration)."

Those are not the same promise.

Stop-losses can be hit by temporary volatility. They can trigger in fast markets and sell you out before the rebound you didn't expect. They also don't guarantee an exact exit price in a gap-down scenario. In a panic, the market can jump past your stop like it wasn't there.

A put's protection is time-defined: it exists for the life of the option. You can hold through noise and still keep your floor. If your objective is to prevent forced selling and preserve decision-making quality during stress, that distinction matters more than most people realize.

The real cost: premium drag and the ``insurance treadmill"

Investors usually underestimate protective puts in two opposite ways: they underestimate how well they protect in a true crash, and they underestimate how painful it can feel to pay for protection in normal years.

The premium has a name for a reason: *premium drag*. It drags on returns in quiet markets because it's a recurring expense. If markets grind higher for years, your puts will often expire worthless. That doesn't mean the strategy "failed." It means the disaster didn't happen. Insurance is supposed to feel slightly annoying in the years you don't use it.

But there's a deeper, more corrosive effect: the insurance treadmill. Each expiration forces a decision:

- Renew, and feel like you're wasting money again.
- Don't renew, and discover the market can smell your confidence.

Many investors buy puts right after volatility spikes because fear is fresh, premiums are high, and the recent past feels permanent. Then markets calm, premiums cheapen, and the investor cancels the hedge because it "seems unnecessary." That is the exact opposite of disciplined hedging. The protective put rewards the investor who can buy protection when it feels boring, not when it feels urgent.

How to think about the premium like an adult

If you want the protective put to be a tool rather than a temptation, reframe the premium in one of these ways:

Premium as a subscription to rational behavior.
If the hedge prevents one catastrophic decision—panic selling, abandoning diversification, derisking at the bottom—it can pay for itself in a way performance charts don't capture.

Premium as a deliberate reduction in equity exposure during a window.
Buying a put is not magically free; it's a way of temporarily shaping your risk profile without selling the underlying. In some situations (taxes, concentrated positions, lockups, or just stubborn long-term conviction), that's exactly what you need.

Premium as a known loss for an unknown benefit.
You're swapping an uncertain tail loss for a small, certain loss. Humans are strangely bad at valuing that trade until they've lived through the tail event.

Practical gotchas that matter in real accounts

The protective put is straightforward, but "straightforward" in options still comes with footnotes that can bite.

Expiration is not a suggestion.
Your floor disappears on a date certain. If you want continuous protection, you must plan the roll: when you'll evaluate new strikes, how far before expiration you'll act, and how you'll avoid turning rolling into a last-minute scramble.

Exercise and assignment mechanics exist even if you never plan to use them.
Even if you intend to sell the put rather than exercise it, you should understand that exercising a put means selling shares at the strike, which requires that you actually have the shares (or can borrow them, which brings its own complexity). In a crisis, operational simplicity is a feature.

Liquidity and spreads.
In calm markets, options can look cheap and easy. In stressed markets, bid/ask spreads can widen and execution matters. A hedge that looks perfect on paper can feel slippery when you try to adjust it under pressure.

None of these issues are reasons to avoid the strategy. They are reminders that "portfolio insurance" is a product with rules, not a vibe.

When a protective put is the right kind of expensive

The protective put shines when you have a reason to stay invested but a reason you can't tolerate deep drawdowns—especially over a defined window. Examples:

- You're carrying a concentrated position you can't (or won't) sell yet.
- You're approaching a financial deadline where sequence-of-returns risk matters: retirement, a home purchase, a business funding need.
- You know your behavioral history: big drawdowns cause you to abandon good plans.

It's less compelling when the objective is "reduce volatility a bit" or "make income." You can sometimes do those things more cheaply with other structures, at the cost of different compromises. The protective put is not a general-purpose accessory; it's a specialized tool for investors who value a hard floor more than they value maximizing average returns.

Actionable takeaway

Write down three numbers before pricing any puts: (1) the date you need protection through, (2) the maximum loss you're willing to tolerate on the hedged position over that window, and (3) the annual premium budget you can stick with even if the market keeps rising. Then choose a put strike and expiration that make those three numbers true in plain arithmetic—and set a calendar reminder to review the hedge *before* it expires, not after.

4.3 Collars: Financing Your Floor with a Ceiling

The protective put is a beautiful invention with one ugly habit: it sends you an insurance bill. A collar is what you do when you like the *floor* but hate the *premium*. You keep the downside protection of a put, but you offset some (or sometimes most) of its cost by selling something you already own: your upside beyond a certain price.

That trade is not subtle. You are exchanging the right to brag about a runaway rally for the right to sleep through a selloff. If the protective put is full-coverage insurance, the collar is the policy with a very specific clause: "We'll protect you—but don't expect to win the lottery."

The collar in one sentence

A *collar* combines three pieces, usually all with the same expiration:

- you own the stock (or ETF),
- you buy a put (the floor),
- you sell a call (the ceiling).

A *call option* gives someone else the right to buy your shares at a fixed price (the call strike). When you *sell* that call, you get paid a premium up front. That premium is the funding source. It's the coupon you clip for agreeing to cap your upside.

So the collar is a tunnel. Below a certain price, the put starts catching you. Above a certain price, the short call starts limiting you. Your returns are nudged into a range that's narrower, calmer, and (crucially) cheaper to maintain than a standalone protective put.

Why investors fall in love with collars (and why they sometimes regret it)

The emotional appeal of a collar is immediate: it feels like you found a loophole. "Wait—I can get protection *and* get paid for it?" In the best case, you can structure a *zero-cost collar* where the premium from the call sale roughly equals the cost of the put purchase. No check written. No premium drag showing up as a visible line item.

But "no check written" does not mean "no cost." The cost is opportunity. If the market rips higher, you don't participate beyond your ceiling. A collar doesn't eliminate the bill; it changes the currency you pay in. Instead of paying dollars today, you pay potential dollars you might have earned tomorrow.

Regret tends to show up in one of two moments:

- the stock rallies hard and you realize you sold away the part that feels best, or
- the stock dips modestly, not catastrophically, and you realize the hedge didn't make you money—it just made you less unhappy.

A collar is happiest in a world where you value stability over heroics, and where you'd genuinely be satisfied earning "enough" rather than "as much as possible." That's not a personality flaw. It's often maturity.

A real-world scenario: the concentrated stock you can't quit

Collars have a natural home in concentrated positions: company stock, a legacy holding with a huge unrealized gain, or a single ETF that's become a dominant slice of your portfolio.

Imagine you own a stock at $100. It's too big, but selling it triggers taxes, or it's restricted, or it carries emotional weight ("I bought this at $12; it paid for my first house"). You want protection, but paying a fat premium for puts every year feels like pouring money into a hole.

A collar offers a compromise that is almost tailor-made for this situation:

- Buy a put with a strike at $90 (you're setting a floor you can live with).
- Sell a call with a strike at $110 (you're agreeing to sell your upside above a level you'd be thrilled with anyway).

If the stock collapses, you have a defined exit via the put. If the stock soars, you may be forced to sell at $110 (or settle the option position in some other way), effectively taking profits at a price you pre-approved. The collar turns "I don't know what I'll do" into "I already decided what I'll do."

This is the collar's quiet strength: it can serve as a *decision pre-commitment device*. In stressful markets, having pre-committed decisions is worth more than cleverness.

How the payoff really works

At expiration, the collar's outcomes tend to fall into three zones:

1) *Below the put strike: losses are limited.* If the stock falls under the put strike, the put rises in value and starts offsetting further losses. Your downside is capped (roughly) at the put strike minus your original stock cost, adjusted for the net option premium.

2) *Between strikes: you behave a lot like an unhedged stockholder.* In the middle zone—between the put strike and the call strike—both options may expire worthless, and you simply own the stock. You'll have some small difference because of the net premium paid or received.

3) *Above the call strike: gains are capped.* If the stock rises above the call strike, the short call works against you. Any gains above the strike are offset by losses on the short call. Your upside is limited (again, adjusted for the net premium).

That "adjusted for the net premium" line matters. If your collar is truly zero-cost, your floor and cap are mostly determined by the strikes. If you pay a net debit, your floor effectively moves down (worse) and your cap effectively moves up (better). If you collect a net credit, your floor improves and your cap tightens. Every collar is an argument about what you're willing to sacrifice.

The moment of truth: ``Who did I sell my upside to?"

Selling a call is not a polite theoretical action. You're taking the other side of a contract. Someone has the right to buy your shares at the strike.

Many investors only notice the reality of that contract when the stock crosses the ceiling and suddenly the position feels less like "income" and more like "handcuffs." You start rooting against your own holding. You flinch at good news. Your stock becomes a source of annoyance.

That emotional twist is part of the price.

If you're considering collars, it's worth rehearsing that moment in advance: *If the stock rallies 25% in three months, will I be angry or satisfied?* If the honest answer is "furious," then the collar isn't a hedge. It's a future self-betrayal.

The collar payoff diagram (the ``tunnel")

The reason collars are so teachable is that the payoff looks like a bracket: losses stop getting worse below the floor, and gains stop getting better above the cap. You can see the compromise in one glance.

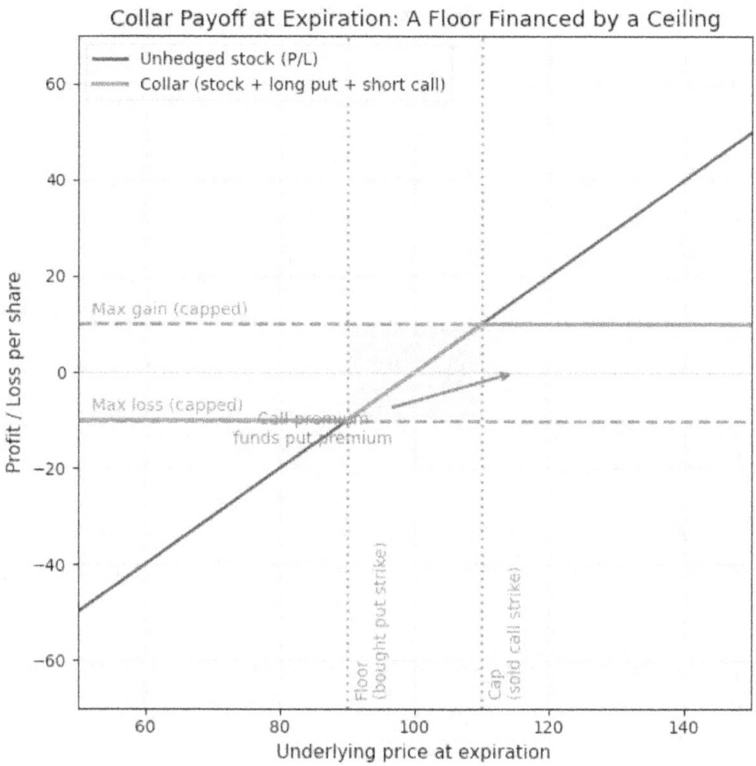

``Zero-cost" is a slogan, not a law of physics

A so-called zero-cost collar is typically constructed by choosing strikes so that the call premium roughly offsets the put premium. But "roughly" hides a lot:

- Option prices move. The quote you like can vanish quickly.
- Bid/ask spreads mean your actual execution can differ from the mid-price.

- Dividends, volatility, and interest rates influence pricing in ways that can shift the balance.

Even if you achieve a perfect zero-cost setup on day one, your *opportunity cost* is still very real. If the stock rockets above the cap, you didn't pay dollars—you paid upside. That payment can feel invisible until it becomes enormous.

So it's healthier to think in terms of "*cost reallocated*" rather than "cost eliminated." A collar is a financing strategy.

The short call: where the elegant diagram gets a little sharp

The collar's clean bracket shape comes from the short call, and the short call is also where the strategy can become operationally annoying.

When you sell a call, you take on *assignment risk*: the possibility that the option buyer exercises their right and you are obligated to sell shares at the strike. For standard U.S. equity options, assignment can happen at any time (they are typically American-style). This can matter if:

- the call is deep in the money,
- there's a dividend approaching (early exercise can be economically rational for the call holder),
- you're holding the collar in a taxable account and don't want a forced sale this year.

None of this is meant to scare you away. It's meant to make the collar feel like what it is: a strategy that trades "premium drag" for "management drag." You may pay less in dollars, but you pay more in attention.

A practical rule: if the collar is supposed to reduce stress, but managing the short call *creates* stress, the strategy is failing its job.

Picking strikes: the art is admitting what you're willing to give up

Strike selection is where investors accidentally reveal their true objectives.

If you set the put strike very close to the current price, you're admitting you can't tolerate much downside. That's fine—it's honest. But to finance that expensive put, you may have to sell a call closer to the current price as well, meaning you're giving away more of your upside more quickly.

If you set the call strike very high because you "don't want to miss out," you may discover the call doesn't raise enough premium to fund meaningful protection. You end up with a collar that's emotionally comforting but financially flimsy: a ceiling you'll probably never hit, paired with a floor that's too low to matter.

A collar forces you to negotiate between two truths:

- You cannot buy a high floor without paying for it.
- Selling upside is a form of payment that feels painless until it isn't.

The most functional collars are often struck at levels where you can say, with a straight face:

> *"If the stock falls to the floor, I'm grateful I had it. If it rises to the cap, I'm genuinely happy to sell there."*

If you can't say both, the collar is probably mispriced for your personality, even if the options are correctly priced by the market.

When collars make the most sense

Collars tend to be at their best in a few specific circumstances:

- *You have a concentrated holding you want to keep, but you need guardrails.* The collar can turn a fragile situation into a bounded one.
- *You have a strong preference for reducing drawdowns over maximizing upside.* The strategy is honest about that preference.
- *You want defined outcomes over a defined window.* Collars are often most useful when tied to a time period that matters (a life event, a vesting schedule, a planned diversification date).

They are less appropriate if your real goal is to capture full equity upside over decades, or if you know you'll resent being called away. The collar is not a free lunch; it's a pre-negotiated compromise.

Actionable takeaway

Before placing a collar, write down two prices: the level where you'd be relieved to have protection (your floor) and the level where you'd be sincerely satisfied to sell (your cap). If you can't name a cap that would still make you happy in a raging bull market, don't finance your hedge by selling calls—either pay for the put outright or hedge a smaller portion so you can keep your upside without resentment.

4.4 Put Spreads: The Deductible Model

A protective put is a heroic promise: "No matter how bad it gets, I have a floor." A collar is a practical bargain: "I'll trade away the moonshot to reduce the bill." A put spread is a third personality entirely: "I'm not trying to survive a meteor strike. I'm trying to stop a normal crash from turning into a personal financial disaster."

It's the deductible mindset. You accept that some pain is part of investing—even *healthy*—but you refuse to let a routine market injury become a permanent scar. You're willing to self-insure the truly extreme tail if it makes the hedge affordable enough to actually stick with.

The basic idea: buy one put, sell another

A *put spread* (specifically, a *bear put spread*) is built from two put options on the same underlying with the same expiration:

- Buy a put with a *higher* strike (this is the put that gives you protection).
- Sell a put with a *lower* strike (this is the put that helps pay for it).

Because you're buying one option and selling another, the trade usually costs less than a standalone protective put. Most of the time it's still a net debit (you pay something), but the cost is reduced because you've sold away the part of the protection that only matters in a truly catastrophic plunge.

That last clause is the entire point. When you sell the lower-strike put, you're saying: "If we go through the floor hard enough to hit *that* level, I'm no longer trying to be fully insured. I can live with the rest."

A story that makes the trade-off feel real

Imagine you own a broad market ETF at $100 and you're not panicking about the end of capitalism. What you're worried about is the kind of year that happens every so often: an ugly 15%–25% drawdown that arrives right when you didn't need the drama. You want protection in that zone. You do *not* want to pay "parachute money" for an "ankle sprain problem."

So you set up a spread:

- Buy the $95 put.
- Sell the $80 put.

Now your hedge has a clear personality:

- From $100 down to $95, you're mostly unprotected (that's your deductible).
- From $95 down to $80, your protection ramps up, dollar-for-dollar.
- Below $80, the protection stops improving. You're still hedged versus the original position, but you no longer gain additional insurance as the market keeps falling.

You've essentially purchased a policy that pays out for the *first* chunk of the accident, not the entire catastrophe.

The ``zone of protection" (and the moment it ends)

This is the cleanest way to visualize a put spread in your head: a put spread buys you a *zone.*

That zone sits between the two strikes.

- The *upper strike* (the put you buy) is where protection begins to matter in a serious way.
- The *lower strike* (the put you sell) is where the protection reaches its maximum and then caps out.

Inside the zone, the spread behaves like real insurance. Outside the zone, it behaves like a reminder that you negotiated.

If you're hedging because you fear "a crash," you should force yourself to put a number on what you mean. Many investors say "crash" when they mean "correction," and pay for protection they don't actually need. The put spread is a tool for investors who can say: *"I care a lot about the first 15%–20%. Past that, I'm not paying retail."*

Why this hedge is cheaper (and what you give up)

A protective put is expensive because it has *convexity* all the way down: the deeper the market falls, the more valuable the protection becomes. That "all the way down" feature is exactly what you surrender with a spread.

When you sell the lower-strike put, you are selling away deep-tail insurance. In return, you reduce premium drag enough that the hedge may become practical to hold repeatedly rather than as a one-time emotional purchase.

The trade-off is not theoretical. If the underlying falls far below the lower strike, your long put continues to gain value, but your short put loses value at nearly the same rate. The two legs start canceling each other out. At that point, the hedge has done what it came to do: it covered the likely damage and then stepped aside.

Put spreads are often the difference between "I should hedge" and "I can hedge." If a protective put feels like buying a designer fireproof suit for a barbecue, a spread feels like buying a fire extinguisher and accepting that you're still not immune to a forest fire.

The arithmetic that matters: max loss and max benefit

Option strategies become less intimidating when you translate them into two numbers you can memorize.

For a typical bear put spread (same expiration):

- *Maximum loss* is the net premium you pay to enter the spread. That's it. It's known on day one.
- *Maximum benefit* is the distance between the strikes minus the net premium paid.

That second line is the "policy limit." If you buy the $95 put and sell the $80 put, the strike width is $15. If the spread costs $4, the maximum benefit is $11 per share (before considering how you're pairing it with your underlying position). Past the lower strike, the spread can't become more valuable than that maximum.

This is not a flaw. It's the structure doing exactly what you asked: *cheap protection with a cap*.

The most common emotional mistake: expecting apocalypse protection at deductible prices

Put spreads are where investors most frequently get angry at math.

They buy a spread because it's affordable. The market drops 35%. Then they stare at the hedge and ask, "Why didn't this save me?"

Because you didn't buy "save me." You bought "help me between these two strikes."

The time to confront this is *before* the trade. Put spreads require you to be honest about your objectives. They're not for investors who need a hard floor regardless of depth. They're for investors who want to reduce the sting of the most common bad outcomes without spending like they're insuring against the rarest ones.

A useful question to ask:

> *If the market drops 50%, will I still be glad I owned this hedge, even though it stops helping below my lower strike?*

If the answer is "no," a spread may not match your temperament, even if it matches your budget.

Choosing strikes: where ``deductible" becomes a decision

Strike selection is where a put spread turns from a generic idea into a personal policy.

Pick the upper strike based on what you consider unacceptable pain. This strike defines where meaningful protection begins. If you're trying to limit a drawdown to something like 10%–15% over a specific window, the upper strike should sit near that threshold (adjusted for the premium you pay).

Pick the lower strike based on what you're willing to self-insure. This is your "I can live with it" line. It might be set at a level you believe would require a true crisis to reach, or simply a level you're not willing to pay to insure.

If you set the lower strike too close to the upper strike, you cap protection quickly and may not cover the drawdown you actually fear. If you set it too far away, the premium savings may be small and you're back to paying up for near-catastrophic coverage.

A put spread is a negotiation between realism and thrift.

A subtle benefit: spreads can reduce ``hedge regret"

Protective puts create a particular kind of psychological whiplash: the market rises, your puts decay, and you feel like you lit money on fire. Collars create a different regret: the market rises *a lot* and you feel like you sold your happiness.

Put spreads can reduce both regrets because they're honest about being *limited*. The premium can be meaningfully lower than a standalone put, and you haven't necessarily put a hard ceiling on your upside the way a collar does.

You still might dislike paying premium in quiet markets—nobody likes paying for insurance—but a smaller premium is easier to live with, and easier to renew without turning the process into a moral crisis.

This is not a guarantee that spreads are "better." It's a recognition that the best hedge is the one you can hold consistently. A theoretically perfect hedge that you abandon after three months is just an expensive hobby.

The short put: where the cheapness comes with responsibility

Selling the lower-strike put isn't just a pricing trick; it creates real-world mechanics you have to respect.

A short put can be assigned. That means you can be required to buy shares at the strike price. Many investors assume assignment only happens at expiration. In practice, early assignment is possible. It's not always likely, but it is possible—and when it happens, it happens on the market's schedule, not yours.

Even without early assignment, expiration can be operationally messy if you hold right up to the final hours:

- prices can whip around the strikes,
- exercise decisions can occur based on small price moves,
- after-hours moves can surprise people who thought "it expired."

The fix is boring but powerful: manage the position *before* it becomes an emergency. Many investors choose to close or roll spreads when there's still time value left, rather than treating expiration like a finish line they must sprint through.

The broader point: put spreads are cheaper partly because you've introduced a short option leg. Short option legs pay you, but they also demand competence and attention.

When put spreads shine

Put spreads are especially useful when your objective is to reduce the damage from the most plausible ugly outcomes:

- You want to protect against a correction-sized decline over a defined window (a year where you're making big contributions, retiring soon, or funding a goal).
- You want hedging that is affordable enough to repeat, rather than a one-off panic purchase.
- You want downside help without explicitly selling away your upside beyond a cap.

They are less appropriate when you genuinely need a hard floor regardless of depth. If your financial life cannot tolerate a 40%–50% drawdown, capped protection can feel like a seatbelt that stops working after the first collision.

Actionable takeaway

If you're considering a put spread, write down two numbers before you look at option chains: (1) the loss level where you want meaningful protection to begin (your upper strike), and (2) the loss level where you're willing to stop paying for insurance (your lower strike). Then check one simple rule: the width between strikes should be large enough that the spread would still meaningfully help in the drawdown you actually fear. If the spread "tops out" before the pain you're trying to avoid, it isn't a hedge—it's a receipt.

4.5 The Covered Call Fallacy

Selling covered calls feels like getting paid to be responsible. The broker deposits premium into your account, your position looks "managed," and the strategy comes with a comforting narrative: you're *doing something* about risk. In a bull market, it can even feel like a cheat code—income on top of gains.

Then a real drawdown shows up, and the illusion breaks in the most insulting way possible: the premium you collected, the thing that felt like protection, barely makes a dent. Your portfolio still falls. Sometimes it falls almost the same as if you'd done nothing. The "hedge" turns out to be more like a coupon.

The fallacy is not that covered calls are useless. The fallacy is thinking they are *downside protection* in the sense that a put (or a put-based structure) is downside protection. Covered calls can be a perfectly rational income and return-shaping tool. They just don't buy you a floor. And when investors confuse "income" with "insurance," they tend to build portfolios that look safe right up until the moment they're needed.

What a covered call actually is

A covered call is simple:

- You own the stock (or ETF).
- You sell a call option on those shares.

By selling the call, you collect premium. In exchange, you give someone else the right to buy your shares at the call strike price. If the stock rallies above that strike, your upside is capped because the short call starts offsetting further gains. If the stock stays below the strike, the call expires and you keep the premium.

So where does the "hedging" claim come from? It comes from the premium. If you collect $2 in call premium, you can tell yourself you have "$2 of downside protection." Technically, that's true in the narrow sense that your breakeven point is $2 lower. But that's not what most people mean when they say hedge. Most people mean: *I want something that kicks in harder when things get worse.*

A covered call doesn't do that. In fact, it does something emotionally perverse: it often works best in the calm, slightly up markets where you least need help, and it offers the least relief in the violent, down markets where you most want it.

The tissue-paper airbag problem

Here's the cleanest way to see the limitation. Imagine you sell a call and collect 2% premium over a month. Great. Then the market drops 12% in two weeks.

What did the hedge do?

It "protected" 2%. You still ate 10%. If the market drops 25%, you still ate 23%. That's not a floor. That's a rounding error with marketing.

In insurance terms, a covered call is like an airbag made of tissue paper: it's technically a safety feature, and it might help in a gentle fender-bender, but it's not what you'd rely on in a highway-speed collision. Yet investors routinely drive into the financial equivalent of a pileup believing the premium will cushion them.

The deeper issue is that a covered call is not *convex*. Convexity is the property you want in true crash protection: the hedge becomes more valuable as losses accelerate. Puts have that. Put spreads have it within their protection zone. Collars have it on the downside because of the put. Covered calls do not. Their benefit is mostly fixed at the premium you collected, and that benefit is dwarfed by large moves down.

A quick reality check with numbers

Suppose you own a $100 stock and sell a one-month call for $2.

Scenario A: The stock falls to $80.

- Stock loss: $20
- Premium collected: $2
- Net loss: $18

Your "hedge" reduced the loss from 20% to 18%. That's not nothing, but it's not portfolio insurance. It's a small buffer.

Scenario B: The stock rises to $120.

- Stock gain: $20
- Short call offsets gains above the strike (depending on strike selection)
- Premium collected: $2

Now the strategy's true nature is obvious: you sold away upside. The payment for your "protection" is that you might not fully participate in the very rallies that make long-term equity investing work.

This is why covered calls can feel emotionally fair in quiet markets and emotionally cruel in raging bull markets: they trade away the exciting part of stocks in exchange for a small, steady-looking premium. The premium is real. The trade-off is also real.

Why the myth persists

The covered call myth survives because the strategy generates *cash flow*. And cash flow is persuasive. A dividend check feels like proof. A premium deposit feels like skill. Investors are wired to trust income because it is tangible.

But options premium is not a dividend. It is not free money. It is compensation for taking on an obligation: you are short volatility (meaning you tend to benefit if things stay calm) and you have limited participation if things go up too much. In simple language: you get paid because you're selling someone else insurance. And the insurance you're selling is on the upside.

The story "I'm hedging because I'm collecting premium" is also psychologically comforting because it reframes risk management as something that pays you rather than something you pay for. Paying for protection feels like admitting vulnerability. Getting paid feels like control. Markets don't care about your feelings. They care about the contracts.

``But covered calls reduce volatility!" Yes---and that's not the same as protection

Covered calls can reduce volatility in a narrow statistical sense. The premium can dampen returns and the cap can reduce upside spikes. In some markets, this can produce a smoother line. A smoother line is pleasant. It is not the same as a floor.

A floor changes the shape of your worst case. It makes "down 50%" become "down 15%" (or whatever you designed). Covered calls don't do that. They turn "down 50%" into "down 48%", give or take the premium and the time period. That's not what most investors are trying to buy when they say "hedge."

It's entirely possible to prefer the covered call profile: fewer blowout months, more income-looking returns, and a different ride. But it should be chosen for those reasons—not because you think it will save you in a crash.

The benchmark that keeps the discussion honest

One reason covered call debates get sloppy is that people argue from anecdotes: "My calls protected me in 2022" or "My buddy got wrecked." A more honest approach is to point to recognized benchmarks that represent systematic covered call strategies.

Cboe, for example, publishes a well-known benchmark called the *BXM Index*, which is designed to represent a covered call (buy-write) approach on the S&P 500: holding the index exposure and selling call options in a rules-based way. The point here isn't that you need to memorize index tickers. It's that covered calls are not some mysterious edge-case tactic. They're a mainstream, studied return profile with known trade-offs: often lower upside capture and only limited help on the downside.

If you want to evaluate covered calls, evaluate them as a long-term return-shaping choice, not as a crash hedge. Benchmarks exist precisely because the strategy is coherent—it just isn't what people imagine when they say "protection."

The ETF wrapper: when a strategy becomes a product

Covered call strategies have become widely available through consumer-facing ETFs that systematically sell calls on an index or a portfolio of stocks. This is convenient. It also makes the fallacy more dangerous, because the product packaging can imply safety.

A covered call ETF can send you frequent distributions, which many investors experience as "income," even though the tax character and economic source can be complex. It can also look stable during some regimes, because the premium cushions small declines and the strategy tends to like sideways markets. Then a fast drawdown arrives and the product behaves like what it is: equity exposure with a small buffer and capped upside.

Add fees, and the trade becomes even more important to understand. Fees don't transform the payoff; they just quietly shave it.

None of this is an indictment. It's a warning label: convenience doesn't change physics.

The crash thought experiment: what happens when markets gap?

The protective put and its cousins have a simple superpower: when panic arrives, their value can rise sharply. That's the moment you want a hedge to become more valuable, not just remain politely positive.

Covered calls do not have that behavior. If the market gaps down overnight, your short call position may actually help slightly (calls lose value when the underlying falls), but the gain is limited to the premium and whatever time value remains. The stock's loss is the main event, and it dominates.

This becomes painfully clear in the kinds of markets that create "down 3%, down 4%, down 5%" days in a row. Premium doesn't compound the way losses can. A covered call buffer is linear; a crash is not.

So if your hedging objective is "catastrophic crash protection," covered calls are simply the wrong tool. They are not a cheaper substitute for puts. They are a different strategy with a different mission.

What covered calls are actually good for

Covered calls have honest uses. They can make sense when:

- You are willing to sell your holding at a pre-chosen price (the strike) and want to get paid for setting that intention.
- You believe returns will be muted or sideways over the option's life and you want to monetize that view.
- You prefer a portfolio that trades some upside for more frequent cash flow.

Notice what's missing: "survive a crash." If you need a floor, you need puts (directly or indirectly) or another structure designed to create convex protection.

The cleanest way to describe covered calls is not "hedged equity." It's "equity with a sold ceiling and a small cash cushion."

The dangerous substitution: swapping insurance for yield

The most hazardous version of the covered call fallacy is when investors replace explicit protection with option income because it feels self-funding. "Why pay premiums when I can collect premiums?" sounds clever until you realize you have flipped from being the insured to being the insurer.

Selling options can be a disciplined strategy. It can also be a stealth way of taking on risk you don't recognize until the bill arrives. In calm markets, the income can feel like genius. In stressed markets, the losses can feel unfair. They're not unfair; they're the contract.

If you want protection, buy protection. If you want income, sell something (upside, volatility, or time) in exchange for it. Mixing those objectives is how portfolios end up with the wrong kind of safety.

Actionable takeaway

Write one sentence about your covered call plan and make it brutally specific: *"I am selling calls to generate income and I accept that this does not create a crash floor; my maximum downside protection is roughly the premium collected."* If that sentence makes you uncomfortable, don't use covered calls as a hedge. Either buy put-based protection for the risk you fear, or reduce exposure directly. The market will happily let you confuse income with insurance—right up until it hurts.

4.6 Hedging with Index Proxies

Buying a hedge on every stock you own is the moment many investors decide they don't really want to hedge.

It starts innocently. You own 25 names. You tell yourself you'll "just buy a put" on the big ones. Then you notice the portfolio risk isn't only in the big ones—it's in the way everything seems to fall together when the market gets scared. You open an options chain for the first ticker, then the second, then the third, and suddenly you're running a tiny derivatives desk in your spare bedroom. Different expirations. Different strikes. Different liquidity. Different implied volatility levels. A calendar full of rolling decisions. It's not just complicated; it's fragile.

Index proxy hedging is the adult alternative: build one fence around the whole herd.

Instead of hedging each sheep, you hedge the pasture.

The proxy idea: hedge the common cause, not every symptom

Most diversified equity portfolios have a big shared driver: the overall market. Individual stocks have their own stories, but in a real selloff the headlines don't care about your stock-picking thesis. Correlations rise. Everything becomes "risk assets." This is why a broad index option can do a surprising amount of work for a multi-stock portfolio.

An *index proxy hedge* uses options on a broad market instrument—commonly the S&P 500 or Nasdaq-100—to reduce the portfolio's exposure to market-wide declines. In practice, that often means:

- buying puts on an index or index ETF, or
- using put-based structures (spreads, collars) on that index exposure.

The hedge won't match your portfolio perfectly, because your portfolio is not the index. That mismatch is called *basis risk*. Basis risk is not a footnote; it's the entire trade. You are swapping precision for simplicity.

If you can tolerate imperfection, you get a hedge you can actually manage.

Why proxy hedging is tempting even for stock pickers

Even investors who love individual stocks often end up admitting something quietly important: they don't really want to hedge *each company*. They want to hedge the *environment*.

If the market drops 20% because of a recession scare, it doesn't matter that you own "quality." It doesn't matter that your companies have "pricing power." In the short run, a broad drawdown can overwhelm individual fundamentals. The objective becomes survival, not elegance.

Index options are designed for exactly that kind of survival math. They're liquid, widely traded, and standardized. In many cases, they're easier to price, easier to execute, and easier to roll than a grab bag of single-name options where one contract trades like water and another trades like a rare stamp.

There's also a psychological benefit: proxy hedging keeps you from overfitting. When you hedge stock-by-stock, you're tempted to express opinions about each name: "This one is safer," "That one is more volatile," "This one needs more protection." You can talk yourself into endless complexity. An index hedge forces humility: "I'm hedging the market. I accept I can't micromanage everything."

SPY vs. SPX: two common doors into the same room

Two widely used S&P 500-related underlyings are:

- *SPY*: an ETF that tracks the S&P 500. Options are equity-style, commonly used by retail investors, and are tied to 100 shares per contract.
- *SPX*: an index option on the S&P 500 itself. These options are cash-settled and are typically European-style (exercise at expiration rather than any time), which can simplify certain operational concerns.

The differences matter less than the principle: you're using a broad instrument to offset broad portfolio risk.

European-style, cash-settled index options can reduce the "call-away" anxiety that comes from short options on single stocks. With some index options, you're not waking up to a surprise assignment that forces you to trade shares you didn't want to trade. That doesn't make them risk-free, but it changes the kind of operational risk you carry.

Meanwhile, SPY options feel more tangible because they map neatly to shares. Many investors prefer that concreteness.

Either way, the key is that the hedge is now one (or a few) positions you can monitor without turning hedging into a second career.

The first uncomfortable truth: your hedge will not be perfect

Proxy hedging is not a magic shield. It's a bet on correlation.

If your portfolio is heavy in tech and you hedge with the S&P 500, you may discover a painful truth: sometimes your portfolio drops more than the index. Sometimes it drops less. Sometimes your stocks get hit by a sector-specific event while the index shrugs. The index hedge can be "working" and you still feel unprotected because your portfolio's drawdown is driven by something else.

This is basis risk in the wild. It shows up in several common ways:

Sector tilt.
If you're overweight one sector (say, semiconductors), your portfolio may behave like a leveraged version of the index on the way down.

Factor exposure.
Value, growth, quality, small cap, momentum—these factors can diverge sharply from the broad market in certain regimes.

Single-name blowups.
An index hedge won't save you from the thing the index doesn't feel. If one of your holdings gaps down 40% on earnings, the S&P 500 put is not your hero.

Proxy hedging is for market risk. It is not a substitute for diversification discipline, position sizing, or the occasional act of admitting you own a landmine.

Sizing the hedge: ``How many puts do I need?" is the wrong question

The right question is: *How much market exposure am I trying to reduce?*

This is where investors often stumble, because they think in tickers rather than in *dollars of risk*. The practical approach is to estimate your portfolio's sensitivity to the index, often described with *beta*.

Beta is a rough measure of how much your portfolio tends to move relative to the market. A beta of 1.0 means "moves like the market." A beta of 1.2 means "tends to move 20% more than the market." It is not a law of nature; it changes when you need it most. But it's a starting point.

Once you have a ballpark beta, you can *beta-weight* your portfolio to the index:

$$\textit{Index-equivalent exposure} \approx \text{Portfolio value} \times \text{Beta}$$

If your portfolio is $500,000 and you estimate beta at 1.1, you might treat it like $550,000 of S&P 500 exposure for hedging purposes. Then you decide what fraction of that exposure you want to protect, and over what range of losses.

This is the practical genius of proxy hedging: you don't need to be perfect. You need to be directionally right enough that the hedge meaningfully changes your worst-case experience.

Choosing the structure: the same building blocks, just scaled up

Once you hedge with an index proxy, you're still choosing among the same basic payoff shapes:

Index protective put (buying the floor).
Most direct. Usually most expensive. Best when you truly want a hard floor for the market-driven portion of your risk.

Index collar (financing the floor).
Lower net premium, but you give up upside on the hedged exposure. This is easier to accept when the goal is "keep my plan intact," not "capture every last dollar of upside."

Index put spread (the deductible model).
Often a sweet spot for long-term investors: meaningful protection through a defined drawdown range, at a cost that doesn't feel like a permanent tax on compounding.

The interesting shift when you use an index is that the hedge feels less personal. You're not hedging *your favorite stock*. You're hedging the weather. That psychological distance can actually make you more disciplined about managing it.

A vivid example: hedging a messy portfolio with one clean position

Say your portfolio looks like a real person's portfolio:

- 12 individual stocks across sectors,
- a dividend ETF,
- a growth ETF,
- a couple of "I know this is risky but I love it" positions.

You could attempt to hedge each piece with its own option. Or you could admit what's actually happening in a broad selloff: your portfolio behaves like "equities plus personality."

Proxy hedging says: protect the equity beta, and accept that idiosyncratic risk (the personality) is something you manage with sizing and diversification, not with an options spiderweb.

So you buy an index put spread that covers, for example, the next 10%–20% market decline over the next six months. Now, if the market gets punched, the hedge responds. Not perfectly, but materially. The point isn't to create a perfectly flat equity curve. The point is to keep a bad season from becoming a plan-ending event.

The second uncomfortable truth: correlations change when you most care

In calm markets, your portfolio might dance to its own music. In crises, the band often plays one song: "Sell risk."

That's why index hedges can be remarkably effective in true panics—not because the hedge is precise, but because correlations rise. The index becomes a blunt but powerful proxy for the market's mood.

But the opposite can happen too. In certain environments, sector-specific shocks dominate. Your portfolio can underperform the index, and your hedge can feel insufficient. The investor who expects a proxy hedge to be a perfect mirror will be disappointed. The investor who expects it to be a stabilizer will often be pleased.

The key is to match the hedge to the objective you wrote down earlier: are you trying to prevent catastrophe, smooth volatility, or limit a shallow drawdown? Proxy hedging can do all three, but only if you accept that "good enough" is sometimes the most professional standard available.

Practical considerations: simplicity has its own risks

Index proxy hedging is simpler, but it's not automatic. A few practical realities matter:

Rolling is still a job.
Protection expires. If you want continuous coverage, you must renew or roll. The simplicity is that you're rolling one structure, not thirty.

Liquidity helps, but execution still matters.
Index options are generally liquid, which is a gift during stressed markets. But spreads can widen when fear spikes. Plan your adjustments when markets are merely nervous, not when they're screaming.

Account constraints are real.
Some accounts restrict certain option strategies. Some in-

vestors don't have approval for spreads. Some are using retirement accounts with limitations. Your perfect proxy hedge is useless if your account can't hold it.

Know what you own.
Index products can have different settlement mechanics. Some are cash-settled, some are share-settled via ETFs. Some are European-style, some are American-style. Those details determine whether you face early assignment risk on short legs, and how the position behaves at expiration. "Index hedge" is a category, not a single instrument.

The goal isn't to memorize every rule. It's to respect that a hedge is a contract with mechanics, not a mood board.

A simple way to start without overcommitting

If you're new to index proxy hedging, the best first step is often to hedge *less than you think you need*. Not because you shouldn't protect yourself, but because the first lesson you need is experiential: how the hedge behaves when markets wobble, how it behaves when they fall, how it decays with time, and how you feel managing it.

Hedge a fraction of the portfolio's estimated index exposure. Choose an expiration that matches a real window you care about. Use a structure whose trade-offs you can emotionally tolerate. Then watch it. Not obsessively—just enough to learn what you actually bought.

The right hedge is not the one that looks perfect at purchase. It's the one you can hold, roll, and live with when markets do what they do.

Actionable takeaway

Estimate your portfolio's market sensitivity in one rough number (beta). Multiply your portfolio value by that beta to get an index-equivalent exposure, then decide what fraction of that exposure you want to hedge and over what drawdown range. Implement the hedge using a single index proxy (SPY or an S&P 500 index option) rather than a dozen single-name options—and write down, in advance, the one scenario where you'll roll or adjust (for example: "roll when 30 days remain" or "roll if the index falls through my upper strike"). The power of proxy hedging is not precision; it's manageability.

Chapter 5

Strategy Design and Maintenance

Many investors treat hedging like buying a fire extinguisher-something you purchase once and stick in a closet. But financial markets are living ecosystems, not static buildings. A hedge that protects $100,000 today might leave you woefully exposed when your portfolio grows to $150,000, or it might silently expire right before the crash hits. This chapter moves beyond the initial purchase to the real work: the ongoing maintenance, sizing, and strategic adjustments that turn a simple trade into a robust, long-term shield.

5.1 Sizing the Hedge: Goldilocks and the Three Put Options

The first time most investors buy a protective put, they buy it the way you buy a fire extinguisher: one, because one feels like the right number. Then the market drops, the portfolio bleeds, the put makes a little money, and the investor discovers a nasty emotional truth: a hedge that doesn't materially change the outcome feels like a broken promise.

That disappointment usually isn't because puts "don't work." It's because sizing is a knife-edge decision. Too small, and the hedge is a placebo-comforting, expensive, and almost irrelevant when it matters. Too large, and you've essentially swapped your portfolio's long-term engine for an insurance policy with a very real premium. The sweet spot is the Goldilocks zone: big enough to change your worst month, small enough that you can afford to keep paying the premium without resenting it.

A useful way to think about hedge sizing is to imagine three investors who all own the same $100,000 stock-heavy portfolio and fear the same scenario: a fast 20% market drop. They each buy one put strategy, but they size it differently. Same storm, three umbrellas:

- *No umbrella* (unhedged): you get wet, but you don't pay for gear.
- *Pocket umbrella* (under-hedged): you pay for gear, but your socks still soak through.
- *Golf umbrella* (fully hedged): you stay dry, but you carry it everywhere and it wasn't cheap.

The goal isn't to win an umbrella contest. The goal is to stay functional-financially and psychologically-through the downpour.

Two numbers that matter: notional and delta

Most portfolios are easy to value and hard to hedge because they aren't one clean ticker. You might own broad index funds, a few individual stocks, and maybe some international exposure. Hedge sizing starts by translating "my portfolio" into a *hedgeable exposure*-an amount of market risk you want the put to respond to.

There are two concepts that sound technical but behave like plain English once you use them:

Notional value is the dollar amount of the thing you're trying to protect. If you own \$100,000 of U.S. equities (directly or via funds), your equity notional is roughly \$100,000.

Delta is the put's "how much it reacts right now" knob. A put with a delta of about -0.50 tends to gain about \$0.50 for every \$1 drop in the underlying *today*, all else equal. A put with a delta of about -0.10 reacts weakly at first-more like a smoke alarm that only starts screaming when the kitchen is already on fire.

The key sizing idea is simple enough to write on a napkin:

$$\text{Hedge ratio (delta-adjusted)} \approx \frac{|\Delta_{\text{puts}}| \times \text{Underlying Notional Hedged}}{\text{Portfolio Equity Notional}}$$

If that ratio is 10%, you have a hedge that's *currently* covering about 10% of your portfolio's small moves. If it's 100%, you're roughly "delta-neutral" on day one: small down moves in the portfolio are mostly offset by gains in the puts (before considering the premium you paid).

There's a trap hiding in the word "currently." Delta changes. A put can become more sensitive as the market falls (that's often what you want in a crash), but it can also lose sensitivity as markets rise. Your hedge is not a brick wall. It's a sliding door.

Still, delta is the best first ruler you have-because it forces you to answer the question investors like to dodge: *How much pro-*

tection do you want when the trouble begins, not when it's already obvious?

The three put options (as a sizing mindset)

The "three put options" here aren't three contracts you must buy; they're three attitudes toward protection-each with a predictable outcome.

1) The placebo put (under-hedged)

A small hedge is seductive because it's affordable. You can tell yourself you're "protected" without sacrificing much upside.

But in a real drawdown, a lightly sized hedge often fails the one test that matters: it doesn't change your decisions. If the market drops 20% and you're still down 17% net, you probably don't feel hedged. You feel mocked-like you paid for a seatbelt that tightens after the crash.

Under-hedging isn't irrational. It's sometimes the correct choice if your true goal is *some* convexity (crash sensitivity) without sacrificing too much return in normal years. The problem is pretending it's disaster insurance. It's not. It's disaster *participation reduction.*

2) The Goldilocks put (material, but survivable)

A properly sized hedge doesn't need to make you whole. It needs to keep you in the game.

For a long-term investor, "in the game" usually means:

- you can keep funding your life without panic-selling,
- you can rebalance into cheaper risk rather than fleeing it,
- you can sleep without checking futures at 2 a.m.

A Goldilocks hedge aims to *cap regret*. It won't eliminate drawdowns, but it can turn a portfolio-threatening event into an unpleasant event.

3) The bunker put (fully hedged)

A near-fully delta-hedged portfolio is a different animal. You may dramatically blunt drawdowns, but you've also made peace with paying an ongoing premium and potentially underperforming in many ordinary markets.

This can be appropriate in narrow windows: you just retired, you're about to fund a large liability, you're concentrated in one stock you can't sell yet, or your personal risk capacity is genuinely low. The mistake is to treat bunker hedging as a casual "portfolio improvement." It's more like changing the engine in your car. The ride is different.

A concrete example: a 20% drop and three hedge sizes

Assume a $100,000 equity portfolio faces a 20% drop (–$20,000). Now imagine three hedge settings based on day-one delta exposure:

- Unhedged: hedge gain $0
- Under-hedged (10% delta): hedge gain roughly $2,000
- Fully hedged (100% delta): hedge gain roughly $20,000

These are deliberately simplified numbers. Real-world results will differ because delta changes as the market moves, implied volatility can spike (which often helps puts), and you paid a premium upfront (a very real cost). But the toy example does something priceless: it shows whether your hedge is capable of moving the needle at all.

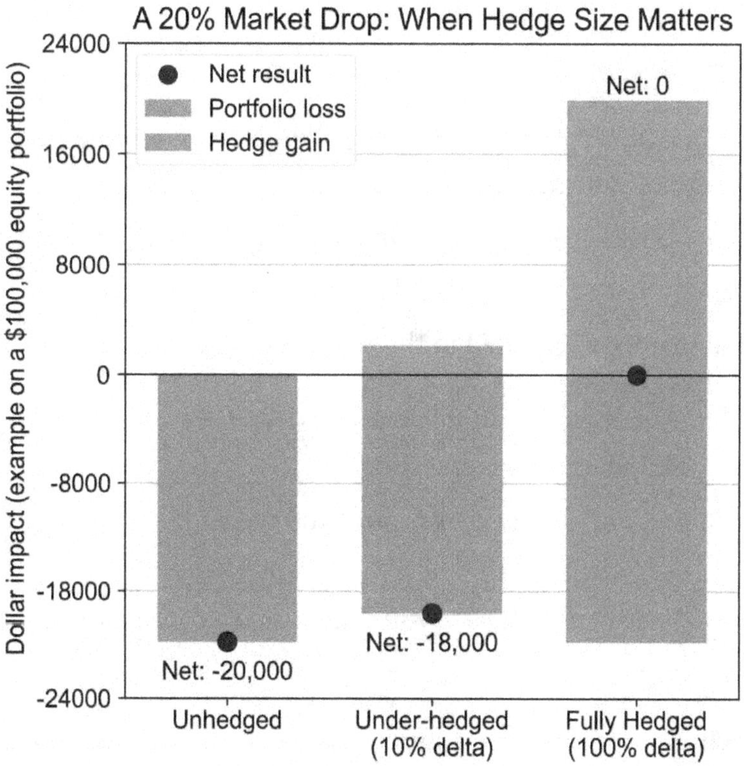

Look at the shape of that picture and notice what it's really saying: the under-hedged investor paid for insurance that only reimbursed a small fraction of the damage. That might still be worth it-but only if they bought it with eyes open. The fully hedged investor "solved" the drawdown (again, ignoring premium), but they also signed up for an experience where their portfolio can feel oddly flat and expensive to maintain in calm years.

The right answer depends on what problem you're trying to solve. And most investors are solving the wrong problem because they never name it.

Name the goal: drawdown cap, survival, or sleep?

Hedges don't just change returns; they change behavior. So instead of starting with contracts, start with a sentence:

"In a fast bear market, I want to limit losses to about ___ so I can ___."

That blank can be filled in three common ways:

- *Drawdown cap*: "I don't want to be down more than 12–15% in a 20% market drop."
- *Survival*: "I need to protect the next 2–3 years of withdrawals so I'm not forced to sell low."
- *Sleep*: "I can tolerate volatility, but I can't tolerate tail risk that makes me abandon my plan."

Each goal points to a different hedge size. Drawdown caps push you toward larger, more expensive hedges. Survival often points to a mix of modest hedging and holding more cash-like assets (because the cheapest hedge is not needing to sell). Sleep can be surprisingly efficient: you might only need enough convexity to keep you from making a catastrophic decision.

If you skip this step, you'll default to vibes-and vibes tend to buy either too little (because premiums sting) or too much (because fear spikes at exactly the wrong time).

A practical sizing workflow (the kind you can do on a notepad)

You can make hedge sizing complicated. Professionals do, because they must. You can also make it *usable*.

Step 1: Estimate what you're actually hedging

If 70% of your portfolio behaves like equities and the rest is bonds/cash, you don't need to buy puts against the whole account.

$$\text{Equity notional} \approx \text{Portfolio value} \times \text{Equity-like fraction}$$

If your $200,000 portfolio is about 70% equity-like, that's roughly $140,000 of equity notional.

Step 2: Choose a starting hedge ratio (delta-adjusted)

For a long-term investor who wants a Goldilocks hedge, a reasonable starting point is often somewhere between "token" and "bunker." In plain language: enough that you'll notice it in a selloff, not so much that premiums dominate your long-run returns.

You can pick an initial range such as 20–50% delta coverage of the equity notional, then pressure-test it against your own reaction. If the market fell 20% and your equity sleeve fell $28,000, would a \$6,000–\$14,000 hedge gain change your behavior? If yes, you're in the territory where hedging becomes more than a story you tell yourself.

Step 3: Translate that into "delta dollars"

Define:

$$\text{Target delta dollars} = \text{Equity notional} \times \text{Target hedge ratio}$$

If equity notional is \$140,000 and your target is 30%, you want about \$42,000 of delta-adjusted protection.

This is the point where people trip because options are quoted in contracts, not in "delta dollars." The bridge is the contract multiplier (typically 100 shares for standard equity options) and the index/ETF level.

Step 4: Convert to contracts (roughly)

Suppose you hedge with puts on an equity ETF at price S. One contract represents about $100 \times S$ dollars of notional exposure.

A single put contract with delta Δ provides roughly:

$$\text{Delta dollars per contract} \approx |\Delta| \times (100 \times S)$$

So the contract count is:

$$\text{Contracts} \approx \frac{\text{Target delta dollars}}{|\Delta| \times (100 \times S)}$$

Input	*Meaning (plain English)*
Portfolio value	Total account size you care about
Equity-like fraction	The part that will likely fall in an equity selloff
Equity notional	Portfolio value × equity-like fraction
Target hedge ratio	How much of equity moves you want to offset *at current deltas*
Target delta dollars	Equity notional × target hedge ratio
ETF/index level S	Price level of what your puts are written on
Put delta $\|\Delta\|$	Sensitivity today; higher means more immediate protection
Delta dollars per contract	$\|\Delta\| \times (100 \times S)$
Estimated contracts	Target delta dollars / delta dollars per contract

This is not perfect, but it's practical. It's the difference between "I think I'm hedged" and "I can estimate the size of my hedge on purpose."

A table you can steal: from portfolio goal to contract count

The following table shows the *logic* of sizing with round numbers. It avoids any specific ticker so you can plug in your own ETF price and put delta.

The most important row is the one most investors ignore: *Target hedge ratio*. If you can't say what ratio you intended, you can't evaluate whether the hedge failed—or whether you simply underbought it.

The hidden cost of "full hedges": you're buying a stream, not a product

A protective put is not a one-time purchase. It's a wasting asset with a clock. That means a fully hedged posture isn't "buy puts once." It's "commit to paying premiums repeatedly."

This is where the Goldilocks problem becomes real life. A hedge that is mathematically perfect but emotionally unaffordable is not a hedge; it's a short-lived experiment.

To keep sizing honest, treat hedging like any other recurring budget. Investors do this naturally with advisory fees and mortgage rates. They oddly refuse to do it with options, even though the options bill can be larger and more persistent.

A useful discipline is to define an *annual premium budget* as a percentage of the equity notional-an amount you can pay without feeling like you're sabotaging your plan. If your budget is 1% per year, then on $140,000 of equity notional you're implicitly saying: "I can spend about $1,400 per year on this form of insurance." If the hedge you want costs 4%, then your real choices are:

- accept the higher ongoing cost,
- reduce hedge size,
- change hedge structure (different strikes/tenors), or
- add non-option defenses (like more cash) so you need less option insurance.

Sizing and affordability are the same decision wearing different clothes.

Why delta sizing beats "number of contracts" thinking

Counting contracts is comforting because it's tangible. But contracts don't map cleanly to protection unless you adjust for:

- the price level of what you're hedging (one contract means different things at different index levels),
- the delta of the put (a far-out-of-the-money put can be cheap because it barely reacts-until it suddenly does),
- the fact your portfolio might not track the hedging instrument perfectly (basis risk).

Delta sizing forces you to confront the present tense: *What does this hedge do for me if the market starts falling tomorrow morning?* Not "what might it do in a cinematic crash," but what it does when the first 5% drop hits and you're deciding whether to hold your nerve.

That present-tense protection is what prevents the most expensive mistake in investing: selling risk at the bottom because you can't take it anymore.

A quick reality check: the hedge that only works in Armageddon

Some investors intentionally buy very low-delta puts because they're cheap and offer explosive payoffs in extreme crashes. That can be a valid *tail hedge*-but it's not the same as a portfolio hedge.

Here's the distinction that saves a lot of regret:

A portfolio hedge helps in bad markets. A tail hedge helps in catastrophic markets.

If your fear is a 20–30% drawdown, but your hedge only pays in a 50% drawdown, you didn't buy protection. You bought a lottery ticket with a thesis.

Tail hedges have a place. They just shouldn't be confused with the thing most long-term investors actually need: a reduction in the kind of drawdown that causes plan failure, forced selling, or behavioral capitulation.

5.2 The Art of the Roll: Avoiding Coverage Gaps

Buying a protective put feels decisive. Rolling it feels like paying rent.

That emotional difference is the first maintenance problem to solve, because the market has a cruel sense of timing: the month your hedge expires is often the month you finally relax. You've watched it decay day after day, you're tired of the premium, nothing bad happened, and the easiest move is to let it die quietly. Then the next week delivers the headline you didn't order.

An option's expiration date is not a feature you can ignore. It's the contract's entire personality. When it expires, your insurance doesn't "fade." It *turns off*. And the most expensive hedging mistake isn't picking the wrong strike—it's discovering you had no hedge on the day you needed one because you drifted into a coverage gap.

Rolling, in one sentence

A *roll* is simply closing your old option position and opening a new one with a later expiration (and often a different strike) so your protection continues without interruption.

Mechanically, it's two trades. Psychologically, it's one decision: *Do I still want to own insurance after seeing the bill?*

If you only ever buy puts once—usually when fear is fresh—you're not running a hedge. You're running a memory.

The calendar risk nobody sees coming

Most investors understand market risk. Fewer appreciate calendar risk: the risk that time passes, contracts expire, and your hedge evaporates on schedule even if your anxiety doesn't.

Calendar risk shows up in three ordinary, unglamorous ways:

- *The "I'll do it tomorrow" gap*: you wait until the final week, life happens, and the put expires before you replace it.
- *The "it's expensive right now" gap*: volatility is elevated, quotes look ugly, so you delay the roll—right when protection is most obviously needed.
- *The "I don't want to realize the loss" gap*: the put has decayed, you hate closing it at a loss, so you avoid the trade and accidentally close your protection instead.

None of these are sophisticated mistakes. That's why they're so common. Rolling is less about mastering options Greek letters and more about mastering the human instinct to postpone painful errands.

The rolling problem is a triangle: time, sensitivity, and price

A hedge has three moving parts that start misbehaving as expiration approaches.

Time decay (often called theta) accelerates as an option gets closer to expiration. That's not a moral judgment from the market; it's just math. The last few weeks are where "wasting asset" stops being a metaphor and becomes a line item.

Sensitivity changes too. Short-dated options can become jumpy: small underlying moves can cause big changes in delta (often called gamma). That can be wonderful when you're correctly positioned, and miserable when you're trying to run a steady, predictable hedge.

Price gets weird in stressed markets. When fear hits, implied volatility often spikes and bid/ask spreads can widen. Rolling during calm markets is like buying insurance from your couch. Rolling during a panic is like trying to buy plywood *after* the hurricane warning.

So the art of the roll is not "always roll on the same date." It's "roll early enough that you're choosing, not being forced."

A simple rolling rule that prevents most disasters

If you want one practical default that prevents coverage gaps without turning your life into a trading desk, use a *time-to-expiration trigger*:

> *Roll when your hedge has about 30–45 days left, and replace it with a new hedge that restores your target maturity (often 3–6 months, depending on your approach).*

Why this works:

- You're far enough from expiration that you're less likely to be trapped by last-minute illiquidity or weird price behavior.
- You reduce the chance that a single chaotic week forces you to roll at the worst possible moment.
- You avoid the "theta cliff" where the option's value melts faster than your patience.

Could you roll earlier? Sure. Could you roll later? Sure. The point is to pick a rule that you can actually follow when you're busy, tired, or emotionally compromised by a market drop.

If you're the sort of person who does oil changes at 5,000 miles instead of "when the engine starts making a sound," you already understand rolling. You're just applying that instinct to risk.

Rolling isn't only about expiration-your hedge's *meaning* drifts

Even if a put still has months left, its usefulness can drift away from what you originally intended. Two common drift problems:

The market moved up. Your put may have become far out-of-the-money. That can be fine if you intentionally bought "catastrophe-only" protection. But if your goal was to cushion a 10–20% drop, a far-out-of-the-money put may not react until the drawdown is already deep.

Your portfolio changed. Contributions, withdrawals, rebalancing, or a big winner can shift your equity notional. A hedge sized correctly three months ago can quietly become an under-hedge without any dramatic event—just the slow creep of numbers.

So a roll isn't merely a calendar event. It's a chance to restore *the hedge you meant to own*, not just extend the one you happen to have.

The two-roll mindset: ``maintenance roll" versus ``crash roll"

Markets hand you two very different rolling environments, and they reward different behaviors.

Maintenance rolls (quiet markets)

In quiet markets, rolling is mostly a budgeting problem. You're renewing insurance when you least feel like you need it. This is where discipline lives, because the market is offering you a psychological discount: no fear, no urgency, no adrenaline—just the invoice.

Quiet-market rolling tends to be the best time to:

- keep your schedule (because you can),
- shop for decent liquidity (because spreads are often tighter),
- restore your intended maturity and delta exposure without pressure.

Crash rolls (fast drawdowns)

In a sharp selloff, your hedge may inflate rapidly. That's the point—but it creates a new problem: what do you do with a hedge that suddenly matters?

A crash roll is less about "renewal" and more about *position management*. You have choices, and none are purely academic:

- *Take profits and re-hedge*: sell part (or all) of the now-valuable puts and buy new protection at a later date/strike. This can lock in gains and keep insurance in force.
- *Let it run*: keep the hedge because you believe the drawdown can deepen. This is emotionally hard because you're watching large, fast P&L swings.
- *Reduce size*: if the hedge has grown large relative to your portfolio (because the portfolio fell and the puts rose), you may be unintentionally over-hedged. That can create a "can't recover" feeling if the market snaps back and your hedge bleeds while your portfolio tries to heal.

A quiet roll is about routine. A crash roll is about not turning a good hedge into a new source of regret.

The coverage-gap trap: waiting for certainty

The most dangerous reason people delay rolling is also the most understandable: they want clarity first.

They want the market to "settle down." They want volatility to "come in." They want the news to "make sense." They want the

Fed to "signal." They want the election to "pass." They want *certainty* so they don't feel foolish paying up for protection.

But the market does not give discounts for waiting politely. If fear is rising, insurance typically becomes more expensive *before* the disaster is officially declared. By the time the risk is undeniable, the hedge you wanted is either pricier, less liquid, or both.

Rolling is one of those rare investing tasks where acting while you feel calm is not just virtuous—it's financially practical.

Execution reality: rolls happen at the bid/ask, not in textbooks

A roll is two trades. Two trades means two spreads, two sets of commissions/fees (if any), and two opportunities for slippage. In calm conditions this is background noise. In stressed conditions it can become a headline inside your own account.

A few execution habits reduce the odds of death-by-friction:

- *Prefer liquid underlyings and expirations.* If you're using options for portfolio insurance, you generally want deep markets with tight spreads. Thin markets are for speculation, not for emergency equipment.
- *Use limit orders.* Market orders in fast markets can turn "rolling" into "donating."
- *Roll as a package when possible.* Many brokers allow multi-leg orders (selling the old put and buying the new put together). This can help you control the net price and reduce legging risk—getting filled on one leg and not the other while prices jump.

Rolling is not only strategy. It's plumbing. And plumbing matters most when the house is shaking.

A roll calendar that fits a human life

A hedge that requires you to be a full-time operator won't survive long in a normal household. The best roll calendar is the one you'll actually follow when you're traveling, working, sick, or simply bored with the whole idea.

One practical approach is to pick a recurring "hedge maintenance day" each month—say the first Monday after monthly statements arrive—and use it to check three things:

- days-to-expiration on the hedge,
- whether the hedge still roughly matches your intended exposure,
- whether your equity notional changed enough to resize on the next roll.

This is deliberately mundane. Rolling should feel like brushing your teeth, not like launching a rocket. You're not trying to outsmart the market; you're trying to avoid being unprotected by accident.

The uncomfortable truth: rolling forces you to admit what a hedge really is

A hedge is not a prediction. It's a subscription.

Every roll is a moment where you face the thing investors hate most: paying for something you hope won't happen. The market has been fine, your put decayed, and now you're asked to buy another one. It feels like getting an extended warranty every quarter—until the day something breaks and you're very glad you had it.

If you size your hedge well, rolling becomes emotionally tolerable because the hedge is doing a job you can articulate. You're not rolling "because options." You're rolling because you made a decision about what kind of drawdown you refuse to tolerate, and the calendar is simply enforcing that decision.

Actionable takeaway: adopt a two-trigger roll rule (time + drift)

Use a roll rule that triggers even when you're busy, even when you're stubborn, and even when markets are weird:

- *Roll Trigger A (Time):* Roll when there are ~30–45 days left to expiration.
- *Roll Trigger B (Drift):* Roll early if the hedge no longer matches the job: portfolio equity notional changed materially, or the put has become so far OTM that it won't respond to the drawdown you fear.

Then add one tiny operational safeguard: set a calendar reminder 60 days before expiration and another at 45 days. The goal is not to be clever. The goal is to never again discover—on the worst possible morning—that your insurance quietly expired last Friday.

5.3 Dynamic Rules: The Thermostat Approach

The most expensive hedge is the one you buy in a hurry, with your heart rate doing the math.

It usually starts the same way: markets are calm for months, your puts quietly decay, and you resent them like a gym membership in February. Then a nasty down week arrives, your news feed turns into an emergency broadcast, and suddenly you want *more* protection—right when it's typically most expensive and least forgiving. That's not a strategy. That's a human being reacting to pain.

A rules-based hedge is the opposite kind of behavior. It doesn't try to predict headlines. It tries to keep your risk at a chosen "temperature," the way a thermostat keeps a house livable. When the room is too cold, the heat turns on. When it's too hot, it turns off. No drama. No speeches. Just a policy.

Why discretion fails: your brain is not a volatility index

The market has a talent for making smart people feel stupid. In quiet periods, buying insurance feels unnecessary; in chaotic periods, *not* buying insurance feels reckless. That emotional flip is the entire trap.

Volatility—how violently prices swing—tends to be mean-reverting. Calm doesn't last forever, and panic doesn't last forever either. Your instincts, unfortunately, tend to do the wrong trade at both ends:

- When volatility is low, hedges feel like wasted money, so you under-hedge or let protection lapse.

- When volatility is high, hedges feel urgent, so you overpay and often over-hedge.

The result is a familiar investing horror: you buy umbrellas *during* the storm and complain about the price of umbrellas *after* the storm.

A thermostat rule is designed to be boring enough that you can follow it—and boring enough that it often works.

The core idea: hedge ratios should respond to *conditions*, not *feelings*

A dynamic hedge policy adjusts your hedge ratio based on observable market conditions, most commonly a volatility measure. The logic is simple:

- When volatility is low, options are often cheaper; you can afford to own more protection.
- When volatility is high, options are often expensive; you either hold what you already own, or you take profits on what has inflated.

This isn't a guarantee. Markets don't sign guarantees. It's just a way to avoid the most common behavioral error: buying high-priced protection after the fear has already arrived.

There's also a mechanical reason volatility matters. As options get closer to expiration, their decay accelerates and their sensitivity to price changes can jump around. Dynamic rules help you resist the temptation to "fix" those changes with constant tinkering. You accept that the hedge is a living thing—and you manage it like one.

You need two dials, not one

A workable thermostat needs at least two dials:

- a *target hedge band*: the range of protection you want to carry (not a single precise number, because markets aren't precise),
- a *trigger signal*: a condition that tells you when to move toward the top of the band or toward the bottom.

The hedge band keeps you from making "all in / all out" decisions that you'll regret. The trigger signal keeps you from drifting into accidental under-hedging or panic over-hedging.

If you skip the band, you'll overreact. If you skip the trigger, you'll procrastinate.

Pick your signal: what counts as ``temperature"?

Volatility can be measured in different ways. You don't need perfection; you need something you can observe consistently.

Common choices include:

- *Implied volatility* (what options prices are implying about future swings). Many investors use a broad "fear gauge" conceptually, because it tends to spike during selloffs.
- *Realized volatility* (what the market actually did over the last 20–60 trading days). This is less flashy and often more stable, but it can react with a delay.

Implied volatility is like the weather forecast. Realized volatility is like the rear-view mirror. A thermostat can work with either, but you should be honest about what you're using: forecast-based rules are more responsive; rear-view rules are less jumpy.

The hidden win isn't which signal you choose. The win is that you choose one *before* you need it.

A simple policy that a real person can actually maintain

Here's a dynamic hedging framework that aims for "practical and survivable," not "quant-perfect."

1) Define a hedge band

Decide on a minimum and maximum hedge ratio (delta-adjusted, relative to your equity notional). For example:

Hedge ratio band: 20% to 60%

This means you never let protection fall below "noticeable," and you never let it grow into "bunker mode" unless you've explicitly chosen that life.

A band also reduces whipsaw. If your trigger signal wiggles around, you don't trade every day. You trade when you're meaningfully out of position.

2) Define volatility regimes

Split the world into a few regimes. Not ten. Not twelve. Enough to be useful, not enough to invite debate.

For example:

- *Low vol*: the market is calm, protection is relatively cheap.
- *Normal vol*: nothing special; maintain the middle of your band.
- *High vol*: fear is elevated; new protection is expensive, existing protection may be valuable.

You can implement this with numeric thresholds if you like, but the deeper idea is what matters: you're pre-committing to behavior that your future self won't enjoy.

3) Connect regimes to actions

A thermostat doesn't ask, "Do I feel like heat today?" It asks, "Is the temperature below the set point?"

Dynamic hedging works the same way:

- In low vol, increase hedge ratio toward the top of your band.
- In normal vol, hold near the midpoint.
- In high vol, avoid impulsively adding; consider taking profits or resizing if you've drifted above your band.

This is where the policy becomes a personality test. When markets are calm, increasing protection feels pessimistic. When markets are scary, not adding feels irresponsible. A thermostat rule is designed to make you mildly uncomfortable at the right times.

Profit-taking is not betrayal: it's part of the job

A surprising source of hedging failure is *success*. When a crash hits and your puts finally light up green, you can become emotionally attached to them. They stop feeling like insurance and start feeling like a hero.

That attachment can create a new problem: your hedge may become oversized relative to your portfolio. Remember, your equity notional just fell. The hedge rose. Even if you never touched the position, your hedge ratio can drift upward fast.

If your original plan was "Goldilocks," a crash can accidentally push you into "bunker." That can be a great problem to have, but you should recognize it as drift, not destiny. A thermostat approach gives you permission to do the sensible thing: trim, roll, or reset so that the hedge continues to serve the portfolio rather than dominate it.

Taking profits on a hedge is emotionally weird because it feels like cheering for bad news. Reframe it. You are not celebrating the fire. You are cashing the insurance check.

The whipsaw problem: when rules become a metronome

Dynamic hedging can go wrong when investors over-engineer it. If you adjust hedge ratios too frequently, you can create a whipsaw:

- volatility drops, you buy protection,
- volatility rises slightly, you sell it,
- volatility drops again, you buy it back,
- and the only consistent winner is transaction costs.

A thermostat works because it has hysteresis: it doesn't flip on and off every time the temperature moves by a tenth of a degree. Your hedge policy needs the same "dead zone."

Two practical ways to build that dead zone:

- *Trade on schedule, not on every tick.* For example, check weekly or monthly, and only act if you're outside your band.
- *Use wide enough thresholds.* If your trigger is too tight, you've built a seismograph, not a thermostat.

The goal is not to perfectly track volatility. The goal is to prevent emotional, expensive extremes.

A dynamic policy still needs guardrails: maturity, liquidity, and simplicity

Even the best trigger signal won't save you from operational mistakes. A thermostat rule is only as good as its guardrails.

Maturity guardrail: Don't let your hedge get so short-dated that it behaves like a live wire. Short-dated options can be powerful, but they can also force frequent decisions and create wild sensitivity changes.

Liquidity guardrail: Dynamic rules can increase trading frequency. If you trade illiquid options, you'll pay for the "dynamics" in spreads. It's hard to run a calm policy through a chaotic execution channel.

Simplicity guardrail: If your rule set requires a spreadsheet that intimidates you, it won't exist when you need it. A hedge policy that depends on your best self is a fragile policy.

A story you'll recognize: the investor who hedges perfectly once

Imagine an investor who buys protection after a scary week, feels proud, and then watches the market rebound. The hedge decays. The investor concludes that hedging is "a drag" and stops.

This isn't an options problem. It's a *policy vacuum*. Without rules, each hedge purchase is judged as if it were a standalone bet. "Did I make money on the put?" becomes the scoreboard.

A thermostat approach changes the scoreboard. The question becomes: *Did my hedge policy keep my risk within the range I chose, at a cost I can tolerate, without requiring heroism?*

If you judge a fire extinguisher by its investment return, you'll throw it away. A hedge policy is the financial version of not doing that.

5.4 When Correlation Breaks Down: The ``Everything" Crash

Diversification is the lie we tell ourselves so we can hold risk with a straight face.

In normal markets, it's a pretty good lie. Your stocks wobble; your bonds behave; your "diversifiers" do their little dances. You feel like you've built a modern financial machine with multiple engines. Then a real panic shows up and you discover the machine has one fuel line. Everything draws from the same tank: liquidity.

Correlation risk is the risk that assets you expected to zig and zag together decide to move as a single crowd. In a chart, it looks like a neat statistic. In a crisis, it looks like your entire portfolio leaning into the punch at the same time.

When that happens, investors often say "correlations go to one." That phrase isn't perfectly literal—markets are messier than a slogan—but it captures the lived experience: the comforting separations between asset classes can shrink right when you need them most. The practical question isn't philosophical. It's operational: *if your diversifiers stop diversifying, does your hedge still work?*

The unpleasant answer is that many "hedges" are only hedges in calm markets. In the everything crash, you find out which protections are structural—built into contracts and cashflows—and which were just statistical friendships that fell apart under stress.

Two crashes, two flavors of ``everything"

The everything crash doesn't always look like 2008. Sometimes it looks like 2020. Sometimes it looks like 2022. The surface story changes, but the plumbing rhyme is the same: forced selling and shifting correlations can turn a carefully diversified portfolio into one big risk trade.

2008 was a credit heart attack. Leverage, banks, and a sudden loss of trust created a brutal feedback loop. Risk assets sank together; safe assets (like high-quality government bonds) tended to help, but even there, the ride was not smooth. The lesson that stuck wasn't subtle: you can't diversify away a systemic funding crisis.

2020 was a different kind of panic. The early phase was a global dash for cash. It wasn't merely "stocks down." It was *liquidity down*. Even high-quality markets experienced stress, and the desire to raise dollars quickly became the dominant trade. When people sell what they can, not what they want, correlations do strange things.

2022 was a quieter, meaner surprise for traditional diversification: inflation pressure and rising rates created an environment where stocks and bonds could fall together. That's not the 2008 story at all. It's not a frozen credit market so much as a repricing of the discount rate across assets. The result felt similar to many investors, though: "my diversifier didn't diversify."

The common thread is that "diversification" is not a permanent property. It's a regime. And regimes change without asking your permission.

Why correlations rise when fear rises

Correlation is often taught like a personality trait: "stocks and bonds are negatively correlated," as if they've signed a lifelong contract. In the real world, correlation is more like crowd behavior. It changes with incentives, constraints, and who is forced to transact.

In calm markets, investors sell based on opinion. In crises, many investors sell based on *necessity*. That shift matters.

Here are the main mechanical reasons correlations tend to rise during panics:

- *Deleveraging is indiscriminate.* If a fund is forced to cut risk, it sells what it owns. It may sell winners to cover losers. It may sell liquid assets because illiquid ones can't be sold quickly. The result is synchronized selling across categories that are "supposed" to be different.
- *Risk models tighten the same belt at the same time.* When volatility spikes, risk-parity and other volatility-targeting strategies may reduce exposures. That's not evil or dumb; it's how they're designed. But it can amplify co-movement because many models respond to the same input.
- *Margin calls don't care about your asset allocation.* If your broker or clearing arrangement demands collateral, you raise cash. And raising cash in a crisis is a blunt instrument.
- *The dollar becomes a gravitational field.* In global stress, the demand for dollars and dollar liquidity can surge. That can ripple through commodities, emerging markets, and even developed markets in ways that make "diversifiers" feel strangely similar.
- *Correlation is not symmetric.* Assets don't always behave the same way in up markets and down markets. The relationships you estimate from average monthly returns can look stable—until you enter the tail of the distribution, where behavior changes.

The phrase "correlations go to one" is shorthand for these forces. Not a law of nature—just a crisis habit.

Calm markets vs crisis markets: what the picture actually looks like

In calm markets, if you plot pairwise correlations between many assets, you tend to see a scatter: some low, some moderate, some negative. The portfolio feels like a committee with arguments.

In crisis markets, the dots often align along a diagonal: correlations rise together, and diversification looks less like a committee and more like a marching band.

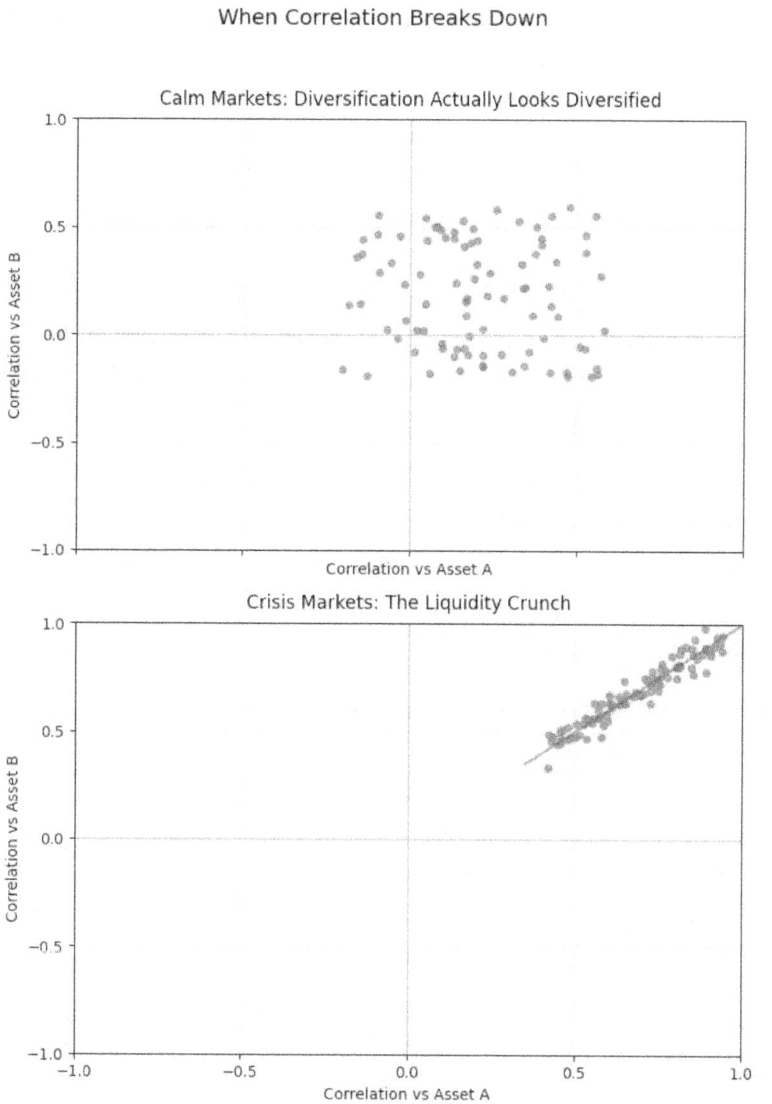

The point of that picture isn't to claim every crisis looks identical. It's to burn a practical intuition into your brain: the relationships you rely on during normal times can compress under stress. If your "hedge" is merely another risky asset you hoped would behave differently, it may not.

Structural hedges vs statistical hedges (and why the difference suddenly matters)

A *statistical hedge* is something that tends to help when stocks fall—based on historical patterns. Traditional examples include high-quality government bonds or certain defensive equity factors. In many regimes, they're wonderful.

A *structural hedge* is something whose payoff is contractually or mechanically linked to the bad outcome you fear. A protective put is the cleanest example: it is designed to increase in value when the underlying drops (with the usual caveats about time decay and volatility pricing).

This distinction is easy to ignore in a bull market because both kinds of hedges can look like unnecessary dead weight. In an everything crash, the difference becomes painfully concrete:

- Statistical hedges can stop hedging when regimes change (inflation shocks are a classic culprit for bond diversification).
- Structural hedges can still work even when markets behave in unprecedented ways—because they are tied to the underlying by design.

This doesn't mean puts are magical. It means they're explicit. They don't depend on a friendship between asset classes. They depend on a contract you purchased.

The liquidity crisis problem: your hedge has to be liquid when *you* are not calm

A hedge that "works" in a backtest but can't be executed in real markets is a philosophical comfort, not a practical defense.

In liquidity stress, three ugly realities tend to appear:

Spreads widen. The bid/ask gap becomes an extra tax right when you're trying to act. Rolling and rebalancing cost more.

Volatility spikes. If you're trying to initiate new protection, premiums can be painfully high. If you already own protection, that spike can be your friend—but it complicates the decision of whether to take profits or keep exposure.

Correlations compress. The assets you expected to sell as "safe" may be down too, or they may be up but illiquid, or they may simply not offset the losses the way your planning assumed.

So "surviving" an everything crash is not only about what your hedge pays at expiration. It's about whether you can maintain, monetize, or resize it under stress without getting trapped by execution costs or decision paralysis.

How to make sure your hedge survives the everything crash

The goal here is not to build a bunker. It's to build a hedge that behaves like equipment: reliable, maintainable, and usable with shaky hands.

Hedge the risk you actually own

If your portfolio is primarily exposed to broad equity beta, hedging a narrow, quirky asset may not help you. In a crisis, quirky bets can become correlated with everything else anyway.

A protective put tied to a broad, liquid equity proxy tends to be a more dependable "systemic stress" tool than a hedge tied to a thinly traded corner of the market. In an everything crash, broad liquidity often matters more than elegance.

Prefer instruments that you can trade in ugly conditions

The liquidity crisis is exactly when you'll want to adjust. That means the best hedges are often the ones with:

- deep option markets,
- tight spreads in normal times (a sign of healthy market structure),
- robust participation (so you're not negotiating with a vacuum).

This is not a call to trade constantly. It's a call to avoid a hedge that becomes untradable when you need it to be tradable.

Avoid relying on a single "diversifier" story

A classic failure mode is overconfidence in one offsetting relationship:

- "Bonds will save me."

- "Gold will save me."
- "International will save me."

Any of these can help in certain regimes. Any of these can disappoint in others. The everything crash is often the regime where disappointment is most costly.

A structural hedge like puts doesn't require you to win the regime lottery. It requires you to pay an ongoing premium and manage the position. That's a tradeoff, not a free lunch—but it's a tradeoff you can understand.

Build a plan for what you do when the hedge is *up big*

This is where many investors freeze. They spent years watching puts decay, then a crash arrives and the hedge finally pays—sometimes dramatically. In that moment, you need rules, because you're holding an asset that is profiting from the pain in the rest of your life.

A practical crash plan answers questions like:

- If the hedge offsets $X\%$ of the drawdown, do you take partial profits?
- If volatility spikes, do you roll to extend protection or do you accept that new insurance is expensive?
- If your hedge ratio has drifted above your chosen band, do you trim to avoid becoming accidentally "short the recovery"?

This isn't about market timing. It's about avoiding two symmetrical errors: refusing to monetize protection when it becomes valuable, and over-monetizing it so you're naked if the crisis deepens.

The emotional trap: the everything crash makes you doubt your own plan

In a genuine liquidity crunch, your screen becomes a betrayal machine. Assets you trusted fall. Your "conservative" sleeve doesn't look conservative. News sounds apocalyptic. And the worst thought appears, quietly and convincingly:

"Maybe the plan was wrong."

That thought is dangerous because it arrives with evidence. Your diversifiers are down. Your correlations are up. Your portfolio is not behaving the way the brochure implied.

This is exactly why explicit hedges exist. A put doesn't fix everything, but it can do something priceless in a panic: it can create *financial breathing room.* With breathing room, you can avoid selling at the wrong time. Without it, you may be forced into the very behavior that turns temporary correlation spikes into permanent losses.

Actionable takeaway: stress-test your diversification with one blunt question

Before the next crisis asks the question for you, ask it yourself:

"If stocks drop 20% in a month, what specifically is supposed to be up, and by how much, and can I trade it under stress?"

If the answer relies on "usually," "historically," or "probably," treat it as a *statistical* hedge and size it with humility. Then decide whether you also want a *structural* hedge—something that is designed to pay when stocks fall, even if the rest of the portfolio starts moving like one big, frightened animal.

In an everything crash, the edge isn't prediction. It's preparedness that still functions when correlations stop keeping their promises.

5.5 Multi-Asset Approaches: The Defense Lasagna

The fantasy version of hedging is a single heroic trade: one put, one moment, one perfectly timed rescue. The real version looks more like a well-packed suitcase. You don't bring one outfit and hope the weather respects your optimism. You pack layers because you don't control the forecast.

A multi-asset defense works the same way. Puts are a powerful layer, but they don't protect against every kind of pain. A fast equity crash is one kind of pain. An inflation shock that drags both stocks and bonds down is another. A liquidity crunch where you suddenly need cash *today* is a third. The "defense lasagna" idea is simply this: build multiple thin layers of protection so that when one layer disappoints, another one shows up wearing a different uniform.

It's also an antidote to a common emotional problem. If you rely on a single hedge, you tend to judge it harshly. When it doesn't pay, you feel foolish; when it does pay, you feel tempted to go all-in on that one thing forever. Layers keep you from turning one market regime into your entire personality.

The first layer is unglamorous: liquidity is a hedge

There's a reason cash feels boring in bull markets: it's supposed to. Cash is the asset that doesn't try to be clever.

But in a bad market, cash becomes the rare investment that can't be gated, halted, or "down 12% on the month." Cash is the only defense that turns volatility into an opportunity by letting you:

- pay bills without selling risk at a discount,
- rebalance into falling assets instead of watching helplessly,
- roll hedges without feeling like every premium dollar is a betrayal.

It's fashionable to say "cash is trash." In an everything crash, cash is oxygen. The investor who can breathe can make decisions. The investor who can't is forced to transact.

A practical way to use cash as a defense layer is to link it to your life rather than your market views. If you withdraw from your portfolio, consider holding a multi-year spending buffer in cash-like instruments. That buffer is not a return engine; it's a behavior stabilizer. It reduces the amount of "portfolio insurance" you need, because it reduces the chance you'll be forced to sell equities in the worst month of your investing life.

Treasuries: the classic shock absorber (with a warning label)

High-quality government bonds-especially longer-duration Treasuries-have historically been the classic shock absorber for equity risk in many deflationary or recessionary scares. When growth expectations collapse and investors rush toward safety, longer Treasuries can rise as yields fall. In those regimes, bonds don't merely diversify; they can actively offset equity losses.

That's the story many investors built their portfolios around, and it was a good story for a long time.

The warning label is that bonds are not a *universal* crash hedge. They are a hedge for a particular kind of crash. When inflation is the villain-when the market is repricing the cost of money itself-stocks and bonds can fall together, and the "shock absorber" starts acting like another weight on the car.

This doesn't make bonds useless. It makes them *regime-dependent*. In the defense lasagna, Treasuries are a layer that tends to defend well in growth scares and deflationary recessions, but may defend poorly in inflation shocks.

That regime dependence is exactly why layering matters. You don't want your entire defense plan riding on a single macro narrative.

Gold: not a crash hedge, but a different kind of insurance

Gold is often discussed with a weird mix of romance and cynicism. Some people treat it like a barbarous relic; others treat it like a truth serum for central banks. A more useful view is simpler: gold is a hedge against certain types of confidence problems.

Gold does not reliably surge every time stocks fall. Sometimes it drops when everything else drops, especially during a "sell what you can" liquidity scramble. If you buy gold expecting it to behave like a put option, you'll be disappointed.

Where gold tends to earn its place is in scenarios where:

- inflation erodes the real value of financial assets,
- real yields fall,
- confidence in currency purchasing power wobbles.

Gold is a layer that answers a different question than equity crash protection. It's less "what if stocks gap down 8%?" and more "what if money itself is being repriced?" In a portfolio built only on stocks and bonds, that's a risk that can hide in plain sight until it's too loud to ignore.

In the lasagna, gold can be a thin layer that doesn't need to be perfect. It just needs to be different.

Options: the structural layer that does what it says on the tin

If cash is oxygen and bonds are a shock absorber, protective puts are the airbag. They are explicit. They are designed to inflate when the underlying drops.

Their weakness is also explicit: you pay for them. Over time, that premium can feel like a leak in the boat. That's why options often work best when paired with other defenses. If you can reduce the size of the option hedge needed-because cash covers near-term spending and bonds/gold cover some macro regimes-then the premium budget becomes more tolerable.

This is the subtle advantage of layering: it can make your most powerful hedge *affordable enough to keep*. The best hedge is the one you don't abandon six months before you need it.

Managed futures / trend: the layer that can change its mind

Most traditional portfolios are long-only. They profit when assets rise and suffer when assets fall. In a crisis, that's a blunt exposure.

Systematic trend-following (often accessed through managed futures strategies) is different because it can, in principle, go long or short across multiple futures markets (equity indices, rates, currencies, commodities) based on price trends. When markets break hard and trends persist, trend strategies may shift exposures in a way that can help during extended drawdowns.

This is not a magic potion. Trend can whipsaw in choppy markets. It can underperform for long stretches. It can feel psychologically irritating because it doesn't "believe" in your favorite assets; it follows price.

But that emotional irritation is partly the point. A defense layer that always agrees with you is usually not a defense layer. Trend is a layer that can respond to crisis dynamics without requiring you to make a discretionary call in the heat of the moment.

In a defense lasagna, trend is often less about being a perfect hedge on day one of a crash and more about potentially helping when the crisis becomes a *regime* rather than a *shock*.

Why layering works: it reduces single-point-of-failure risk

A one-hedge portfolio has a hidden fragility: if your hedge fails in the regime you get, you have no backup.

Layering is not about finding five things that all hedge equities. It's about defending against multiple failure modes:

- *Equity gap risk* (puts are built for this).
- *Deflationary recession risk* (Treasuries often help here).
- *Inflation / currency repricing risk* (gold can be relevant).
- *Liquidity / forced-selling risk* (cash helps by changing what you're forced to do).
- *Prolonged trend drawdowns across assets* (trend-following may contribute in some regimes).

Notice the shift. The question isn't "what's uncorrelated?" The question is "what breaks in each scenario, and what survives?"

This is also why the lasagna metaphor matters. You don't want one two-inch slab of cheese. You want multiple thin layers that together create resilience. Thin layers are easier to keep, easier to rebalance, and less likely to become a portfolio-killing obsession.

A realistic blueprint: how layers interact in the real world

A layered defense has an internal logic that should feel almost physical:

Cash reduces urgency. If you have cash reserves, you don't have to sell equities in a drawdown, and you can maintain option hedges without feeling like you're lighting money on fire.

Bonds can offset growth scares. When they work, they can reduce the depth of the equity drawdown you're trying to insure, which can reduce how "bunker-like" your put hedge needs to be.

Gold can hedge a different enemy. If inflation is the shock, gold may help in ways bonds do not.

Puts protect the left tail. They are the layer that doesn't ask the macro question first. If the underlying drops, they respond.

Trend can adapt. It may not catch the first punch, but it can sometimes reduce exposure as trends persist, potentially helping when the environment stays hostile longer than your nerves would like.

The important part is not that each layer always works. The important part is that they don't all rely on the same reason to work.

The cost of lasagna: complexity calories

Layering can fail for a different reason: it can become a junk drawer.

If you pile on defenses without a plan, you can end up with:

- overlapping exposures that don't actually diversify,
- too many moving parts to maintain,
- a portfolio that is "defensive" in name but random in behavior.

A layered defense needs constraints. Two constraints keep it from turning into chaos:

Constraint 1: each layer must have a job description. You should be able to say, in one sentence, what scenario each layer is meant to help with.

Constraint 2: each layer must have a maintenance plan. If you can't realistically rebalance it, roll it, or hold it through discomfort, it's not a layer-it's a future regret.

How to size layers without turning your portfolio into an insurance company

Layer sizing is where investors often overthink. You don't need precision; you need balance.

A practical approach is to start with *risk budget*, not capital:

- How much drawdown can you tolerate without changing behavior?
- How much premium (explicit or implicit) can you pay for defense?
- How much complexity can you manage without abandoning the plan?

Then allocate defensive responsibilities:

- Use *cash* to cover near-term obligations and reduce forced selling.
- Use *bonds* as a diversifier, but assume they can fail in some regimes.
- Use *gold* (if you choose) as a different regime hedge, not as a put substitute.

- Use *puts* for explicit crash convexity, sized so the premium doesn't make you quit.
- Use *trend* (if you choose) as an adaptive diversifier, sized so you can live with its weirdness.

The best layered defense is one where no single layer has to be heroic. Heroic layers are expensive, psychologically demanding, and hard to maintain. Modest layers can be kept.

A small but powerful reframe: ``defense" is what lets you take offense

The point of protection isn't to become a professional worrier. It's to become an investor who can act when others can't.

If your defenses are doing their job, a drawdown stops being purely a threat and becomes partially an opportunity: you can rebalance, deploy cash, roll hedges deliberately, and avoid the career-ending mistake of selling your long-term assets at the bottom because you ran out of emotional and financial runway.

Defense isn't pessimism. It's the practical foundation that allows optimism to survive contact with reality.

Chapter 6

Execution and Operational Reality

You've designed the perfect crash helmet, but do you know how to fasten the strap? In theory, hedging is about negative correlations and Greeks. In practice, it's about order types, cash management, and surviving the moments when the market plumbing starts to leak. This chapter moves from the whiteboard to the trading floor, exploring the unglamorous but critical machinery of hedging. We will ensure you don't get stopped out by a margin clerk right before your hedge pays off, and we'll teach you how to spot the difference between a safe insurance policy and a hollow promise.

6.1 The Price Tag You Don't See: Mechanics and Liquidity

A hedge that looks cheap on paper can turn expensive at the exact moment you need it. Not because your thesis was wrong, but because trading is not a math problem that settles at "the price." Trading is a negotiation inside a moving crowd. And when the crowd gets nervous, the fee you pay isn't a line item your broker prints for you—it's embedded in the microstructure: the bid-ask spread, the depth behind those quotes, and the slippage that appears when you try to act quickly.

Most investors start with the *last price*. It's the number that flashes on a screen, the one that makes you feel like you know where things are. But the last price is a historical fact, not a guarantee. It's simply the most recent trade someone else managed to execute. Your trade will happen at the *bid* (if you're selling) or the *ask* (if you're buying), and those two prices are rarely the same. That gap is the spread. If you remember only one thing about execution, make it this: the spread is the market's tollbooth, and every round-trip through it costs money.

Think about a protective put—classic portfolio insurance. You decide you want protection, you hit "buy," and you're done, right? Not quite. You pay the ask to enter. If you later sell that put (maybe because the market calmed down, or you're rolling to a new expiration), you receive the bid. Even if the option's "mid" price hasn't moved at all, you can lose money purely because you crossed the spread twice. That's not a theoretical nuisance. In calm markets it might be a paper cut. In a panic, it can be a serious wound.

The spread exists for a reason. Liquidity providers (market makers and other participants willing to stand ready to buy and sell) aren't running a charity. They're taking the risk that, the moment they buy from you, the market drops; or the moment they sell to you, the market spikes. The spread is compensation for providing immediacy and absorbing that uncertainty. When volatility rises, that risk rises. When news risk is high, that risk rises. When the underlying market is jumpy and fragmented, that risk rises. Spreads widen not because "the market is unfair," but because the insurance company just heard thunder and is raising premiums.

A simple way to feel the spread in your bones is to imagine you're buying a used car. The dealer will sell you one for $20,000, but will only buy it back from you for $18,500. Nothing about the car has changed in the last thirty seconds. That $1,500 is the spread. In markets, the numbers are smaller in percentage terms (usually), but the principle is identical. When you trade, you are stepping into a two-sided quote where someone else has set a buyback price that is *worse than* the purchase price. That difference is your immediate, guaranteed cost.

Now add the second ingredient: *depth*. The best bid and the best ask are only the first step on the ladder. Behind them sits an order book: many more buyers willing to pay a little less, and many more sellers demanding a little more. Depth is how much you can trade without moving the price too far. When depth is plentiful, you can place a reasonably sized order and get filled close to where you expected. When depth is thin, your order has to "walk the book," chewing through the available shares or contracts at the top levels and then spilling into worse prices. The market didn't "change its mind" about the asset; you just pushed into the part of the book where the prices are uglier.

This is where hedging gets operationally real. A long-term investor can sometimes ignore small execution costs when buying a broad index fund once a month. A hedge is different. Hedges are often added or adjusted at emotionally charged moments: after a scary headline, during a drawdown, or when volatility is already elevated. That's precisely when spreads widen and depth evaporates. So the decision to hedge tends to coincide with the moment hedging is most expensive to implement. It's like trying to buy flood insurance after the water has already reached the porch.

The mechanics matter even more with options, because options are naturally "spready." Unlike a mega-cap stock that trades constantly, many option strikes and expirations trade intermittently. You can pull up an options chain and see dozens of contracts with quotes, but some of those quotes are more like suggestions than firm, eager interest. The contract you want might have a wide spread, and the size available at the best bid and ask might be tiny. If you need ten contracts and only two are offered at the best ask, the rest of your order may fill at higher prices. Congratulations: you just paid a hidden surcharge for urgency.

The cleanest mental model is this: there are two prices for any tradable instrument. There's the *price you can brag about* (the midpoint, the last trade, the theoretical value), and there's the *price you can actually get* in the size you actually need, right now, with other people also trying to do the same thing. Hedging lives and dies on the second one.

Market orders are the accelerant that turns these frictions into real damage. A market order is not an order to trade at "the market price." It's an order that says, "I don't care what price I get, just fill me." In a quiet market with tight spreads and thick depth, that might work fine. In a fast market, it's like shouting "I'll pay anything!" in an auction hall where the lights are flickering. You may get filled quickly, but you may not like where you land.

The danger isn't just that you pay the spread. The danger is *slippage*: the difference between the price you thought you'd get and the price you actually get. Slippage is what happens when the quote you saw is gone by the time your order arrives, or when your order is large enough to reach into worse levels, or when volatility is high enough that the book constantly reshapes itself. Slippage is the tax you pay for being late to the party, for needing immediacy, or for underestimating how fragile liquidity can be.

Even limit orders—the "safer" cousin—have their own reality. A limit order says, "Only fill me at this price or better." That protects you from a disaster fill. But it introduces a different kind of risk: you may not get filled at all. In hedging, *non-execution* can be its own cost. If the market is falling and you place a limit order to buy puts at a "reasonable" price, you might watch the protection run away from you while your order sits politely on the sidelines, unfilled. The market didn't cheat you; you tried to negotiate while the room was on fire.

Liquidity also changes shape depending on the product. Large, heavily traded index ETFs and their near-the-money options usually have better liquidity than single-stock options in smaller names. The difference is not subtle in a crisis. If you hedge a diversified stock portfolio with a liquid index product, you're typically leaning on one of the deepest pools of trading interest. If you hedge each individual position with its own options, you're stepping into many smaller pools—some of which can become puddles when volatility spikes. The end result can be that your "precise" hedge is harder to enter and harder to exit than a cruder index hedge.

There's a psychological trap here: when people see a wide spread, they often treat it as an insult. They feel the market maker is "charging too much." But spreads are signals. A wide spread is the market saying, "I'm not confident about the next few seconds, and I'm not eager to warehouse your risk." It's a real-time stress test of how much the market wants to hold inventory. In calm times, liquidity providers compete aggressively and the spread tightens. In anxious times, they protect themselves. The spread isn't a moral failing; it's a weather report.

Depth is the same kind of signal, just quieter. You can look at the top of the book and see a tight spread, feel reassured, and still get surprised because there's no depth behind it. Two cents wide, but only 50 shares offered. Or an option that's a nickel wide but only one contract available. That's the financial equivalent of a bridge that looks solid until you drive a truck onto it. A hedge is often sized in meaningful notional terms. The order sizes involved can turn "looks liquid" into "isn't liquid" very quickly.

A particularly nasty feature of liquidity is that it is state-dependent. When markets are rising and boredom is high, liquidity is abundant, spreads compress, and everyone feels clever. When markets gap down and fear shows up, liquidity tends to *withdraw*. Quotes widen. Size disappears. The visible book thins out. Even instruments considered "safe" can get jumpy if everyone wants the same exit at once. This is why execution is not an afterthought to hedging; it's part of the hedge itself. If your plan requires trading quickly during stress, the plan implicitly assumes liquidity will be there. That assumption deserves skepticism.

The "hidden price tag" shows up not just when you enter, but when you maintain a hedge. Rolling options, adjusting deltas, taking profits, resetting strikes—all of that is trading. If you run a strategy that requires frequent tinkering, you're paying the spread repeatedly. Sometimes that's worth it, especially if the hedge materially reduces tail risk. But you should treat those repeated spreads like recurring premiums, not as harmless friction. A hedge that looks inexpensive because you modeled only option premiums can be materially more expensive once you account for trading costs over time.

A practical way to think about this cost is a simple round-trip estimate. Suppose an option has a bid of $1.90 and an ask of $2.10. The midpoint is $2.00, which looks tidy. If you buy at $2.10 and later sell at $1.90, you just paid $0.20 per contract, ignoring commissions and fees. That's $20 per contract (since options are typically on 100 shares), and $200 if you traded ten contracts. If you roll monthly and do this repeatedly, it becomes a meaningful line item—even if every trade looks "small."

Spreads can also distort your perception of performance. Many platforms mark your position at the midpoint. That makes your P&L look nicer than what you could actually realize. In a wide-spread product, your position can show a profit at mid even though you would lose money if you hit the bid to exit. This is not a conspiracy; it's just how marks work. But it can trick you into thinking your hedge is "working" when the executable reality is less generous.

The order book has another subtle lesson: your hedge competes with other people's urgency. When markets lurch, a lot of investors have the same thought at the same time: "I should protect myself." That collective impulse compresses time. You're not alone trying to buy puts or volatility exposure; you're part of a stampede. Liquidity providers respond by widening spreads and rationing size. The market becomes a place where immediacy is expensive. And a hedge, by definition, often demands immediacy.

None of this means hedging is futile. It means execution is a design constraint. You don't choose between "hedge" and "no hedge." You choose between different ways of paying for protection: explicit premiums, implicit spreads, and the risk of not being able to transact when everyone is panicking. A well-designed hedge respects the fact that the market's plumbing matters. It uses instruments with resilient liquidity, sizes trades to fit the available depth, and avoids turning a protection plan into a series of frantic market orders.

The real investor advantage here is not speed. It's *patience* and *planning*. If you know you want crash protection for the next year, the best time to put it on is rarely the morning after a terrifying headline. The best time is often when you can afford to be picky: when spreads are tight, depth is available, and you can work limit orders without feeling like you're racing a falling elevator. You're not trying to "time the market." You're trying to buy insurance when the insurance desk isn't overwhelmed.

You can also borrow a habit from professional traders: think in terms of *implementation shortfall*—the gap between a hypothetical paper price and what you actually achieved. Even without fancy analytics, you can approximate it. Record the midpoint when you decide to trade, the price you actually get, and the spread at the time. Do this for a few months and you'll start to see which products are quietly taxing you and which ones are operationally friendly. Many investors obsess over management fees measured in basis points and ignore spreads that can dwarf those fees when trading options.

At a nuts-and-bolts level, there are a few execution behaviors that separate "I bought a hedge" from "I implemented a hedge well." Prefer limit orders over market orders for options and less-liquid instruments, especially during volatile periods. Break larger trades into smaller ones if the displayed size is thin, rather than forcing the order to walk the book. Favor liquid strikes and expirations when practical; the perfect strike that barely trades can be more expensive than the slightly imperfect strike with tight execution. And if you're trading around major news events (earnings for single stocks, macro releases for indexes), assume spreads will widen and fills will be less friendly.

One more operational reality: liquidity has a personality. It's often most generous when you don't need it. That's not cynicism; it's the nature of risk. In quiet markets, providing liquidity is relatively safe, so more participants compete to do it. In stressed markets, providing liquidity is dangerous, so fewer participants step up, and those who do demand more compensation. When you design a hedge, you're designing not only a payoff diagram, but also a plan for how you'll behave when the market's personality changes.

That's why execution belongs in the same mental bucket as diversification and position sizing. A hedge that cannot be executed at reasonable cost is like a fire escape blocked by storage boxes: it looks great in the brochure, but it fails when you actually need it. The goal is not perfection. The goal is to avoid avoidable mistakes that turn protection into self-inflicted damage.

Actionable takeaway: Before placing any hedge trade, take 30 seconds to do a "spread-and-depth check." Look at the bid, the ask, and the quoted size at each. Estimate your round-trip spread cost (ask minus bid, times your contract/share quantity). If that number feels like an unpleasant surprise, don't negotiate with reality by hoping for a better fill—change *something concrete*: use a limit order near the midpoint, reduce size, choose a more liquid strike/expiration, or switch to a more liquid hedging instrument.

6.2 Mind the Gap: Crisis Liquidity Risks

Friday afternoon, everything looks merely unpleasant. Your stock closes at $100. You place a stop-loss at $95 because you're disciplined, you're calm, and you refuse to be the person who "rides it down." You sleep like someone who believes the market moves one tick at a time.

Monday morning, the screen prints $80.

Not traded down through $99, $98, $97, politely pausing at your stop. Just *there*: $80, like the elevator cable snapped over the weekend. The stop-loss that felt like a seatbelt turns out to be a note taped to the dashboard saying, "Please brake."

That's gap risk: the market's habit of moving discontinuously, leaping from one price to another with nothing in between. The hidden danger is not the size of the move—it's the *missing prices*. A stop-loss order can only trigger if the market actually trades at or through your stop. In a gap, there may be no trades anywhere near your level. Your "exit plan" is still valid in spirit, but the mechanism you chose to enforce it is suddenly powerless.

A stop-loss is not a guaranteed price. It's an instruction: "If the market touches this level, convert my order into a market order." That conversion is exactly what makes stops feel comforting—until you notice the fine print. The moment your stop triggers, you become the most desperate seller in the room. You've told your broker you don't care about price anymore. In a normal tape, that urgency costs you a little slippage. In a crisis tape, it can cost you the difference between "managed loss" and "story you tell for a decade."

Stop-losses fail in two distinct ways, and both are operational, not psychological. First, a gap can jump cleanly past your stop price, so the order triggers only after the market has already repriced dramatically lower. Second, even without a gap, a fast cascade can create a brief vacuum where bids evaporate, your stop triggers into a market sell, and you get filled at a price that feels like a typo. The market didn't violate the rules; you gave it permission to execute you at whatever bid survived.

The real cruelty is that stop-losses often behave *best* in the environments where you need them least. In gentle, liquid markets they trigger near the level you chose, making them look like a reliable technology. That success builds confidence. Then a genuine stress event arrives—the kind where spreads widen, depth thins, and liquidity providers step back—and the tool finally reveals what it always was: a conditional market order, not a protective floor.

Here's what a weekend gap looks like in the only language that matters: a chart with empty space where your plan was supposed to work.

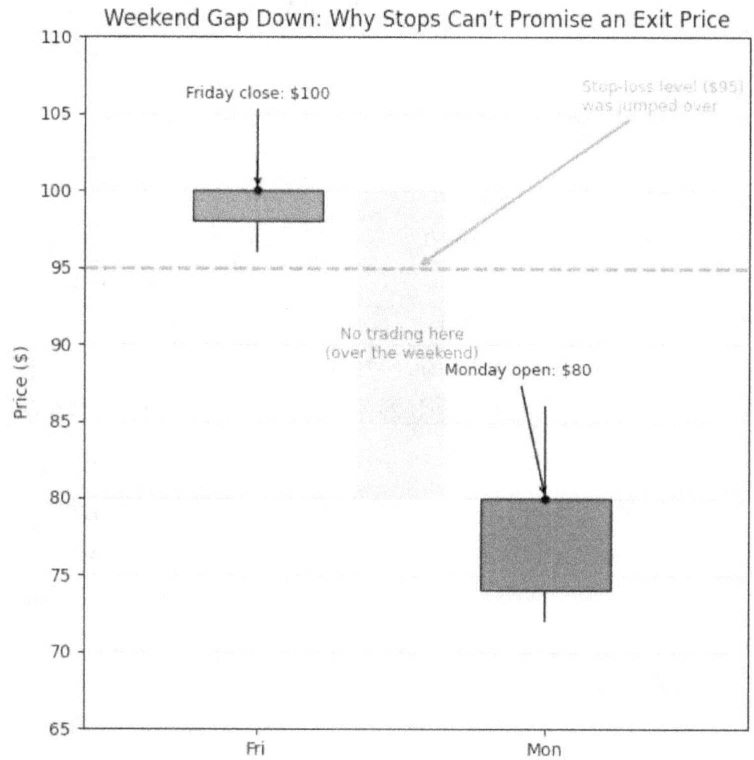

The empty space is the whole point. A stop-loss can't execute at $95 if the market never offers $95. When trading resumes, your order becomes a market sell into a thinner book at a worse moment. You don't "lose 5%." You lose whatever the first real bids will pay. The market doesn't owe you continuity.

Gap risk isn't rare because markets enjoy drama; it's common because information arrives in lumps. Earnings reports drop after the close. Regulators make weekend announcements. Geopolitics doesn't check the futures session before misbehaving. Even when the headline hits during market hours, liquidity can vanish faster than most investors imagine. When everyone wants the same door at once, the "price" becomes less like a number and more like a negotiation with whoever is still willing to stand there.

There's a second, sneakier version of the same problem: you can experience a gap *without* a calendar gap. In a fast selloff, the market can trade $100, then $98, then $95, and then suddenly the next available bid is $90 because the bids in between were pulled. To the chart, it looks like a quick move. To your stop order, it's the same story: your stop triggers and then executes wherever liquidity survived. The missing prices aren't drawn as empty space—they're missing because the book blinked.

This is where the earlier reality about liquidity becomes personal. Spreads and depth aren't just "trading costs." In a crisis, they become *outcomes*. A widening spread is the market telling you it's unsure. A collapsing depth is the market telling you it's afraid. When those signals flash red, a stop-loss is like insisting on an orderly queue during a fire drill. You might get out, but you won't get out *politely*.

A lot of investors respond by upgrading to a stop-limit order. That sounds like a solution: you set a stop at $95, and a limit at, say, $94.50, so you won't get filled at $80 by mistake. The catch is brutal and simple: if the market gaps below your limit, you don't sell at all. You remain fully exposed while the market is falling, which is exactly when you were trying to reduce exposure. Stop-limit orders can protect you from an awful fill, but they can also protect you from *any* fill. In a slow market, that trade-off may be reasonable. In a crash, it can be a trapdoor.

The uncomfortable truth is that stop orders are not *protection*. They are *automation*. They help you follow a rule when your emotions might otherwise sabotage you. That's valuable. But it's not the same thing as a contractual floor. When someone says, "I'm protected; I have stops," what they often mean is, "I have an intention to sell if things get bad." A crisis is the market's way of asking whether you wanted an intention or a guarantee.

Options are one of the few tools that can provide a real guarantee on price, because an option is a contract, not a request. A put option gives you the right (not the obligation) to sell at a specified strike price before expiration. If you own a put with a $95 strike, the market can open at $80 and you still have a mechanism to realize value tied to that $95 strike. You may not get exactly $95 in the market if you sell the option, because option pricing includes time value and implied volatility. But the key difference is structural: the put's payoff does not require the market to "trade through" your level. The contract is designed to *be valuable* when the underlying is below the strike, even if the move was a jump.

That doesn't mean options are magic. They can be expensive. In a panic, implied volatility often spikes, which inflates option premiums. Liquidity in some strikes can still be poor. But the nature of the risk is different. A stop-loss is an attempt to transact in the future under unknown conditions. A put is prepaid insurance: you pay now to avoid being forced to transact later at whatever price the market offers you.

Crisis liquidity risk also shows up in places that feel "safe," which is why it catches sophisticated investors off guard. In March 2020, the world briefly learned that liquidity is not an honorific title—it's a state of the system. Even highly traded markets can seize when leverage, funding, and risk limits collide. When people need cash, they sell what they can, not what they want. That's how you can see strange dislocations: prices moving in ways that don't match the tidy relationships you expect. The implication for hedging is not that you should fear every instrument equally; it's that you should respect the possibility that the exit door narrows precisely when the building shakes.

One of the most dangerous moments is the "overnight reconsideration." During the day, participants can gradually adjust. Overnight, the market comes back with a new consensus and no patience for your timeline. That's when investors discover the difference between *risk you can trade* and *risk you can only endure*. A long-only portfolio can endure a lot if you have time and cash flow. A leveraged or tightly risk-managed portfolio often can't. Stops become more common in those portfolios because discipline is necessary—and that discipline, paradoxically, can concentrate selling pressure at the same levels, making the liquidity problem worse.

A practical example: imagine a widely held stock with bad weekend news. Thousands of investors set stops around the same obvious technical levels: "If it breaks $95, I'm out." Monday opens at $80. Now the stop orders trigger en masse, all converting into market sells at the open. The first prints are ugly, and the next prints can be uglier, because the stop machinery itself becomes a source of forced selling. You wanted a personal safety mechanism; you joined a crowd.

Index-level gaps can be less frequent than single-stock gaps, but they matter more to hedging because many hedges are index-based. Futures markets trade nearly around the clock, but that doesn't eliminate gaps; it simply moves the discontinuity to moments of thin participation. There are hours when the market is "open" yet functionally sleepy, and a sudden burst of news can still produce a sharp repricing with very little depth. If your risk plan assumes you can always adjust at the moment you feel like it, you're assuming continuous liquidity. That's not how real stress behaves.

So what does a long-term investor do with this? The answer is not "never use stops." Stops can be perfectly sensible for certain positions, especially smaller, more speculative holdings where you want a clean rule to prevent a small mistake from becoming a big one. The answer is to stop treating stops as insurance. Use them as a behavioral tool, and buy actual insurance when you genuinely need a floor.

The most robust approach is to decide, *in advance,* which risks you refuse to carry through a discontinuity. If a 20% overnight drop would change your life, that's not a place for wishful execution. That's where you consider hedges that do not depend on you getting a clean fill at a clean time: protective puts, collars, or other structures that define your worst-case with a contract. If an overnight drop would be unpleasant but survivable, you might accept that your "plan" is to ride it out—and then you size your positions so riding it out is actually survivable, not merely aspirational.

There's also a subtle but powerful operational habit: separate your desire for protection from your desire to trade. Many investors place stops because it feels like doing something without paying a premium. But a stop doesn't remove risk for free; it shifts the moment you pay for it, often into the worst possible market. Insurance feels expensive because you pay when the sun is shining. Crisis execution feels cheaper because it's invisible until it isn't. The bill arrives later, when you're least equipped to negotiate.

Even if you do decide to use stops, treat the open like a different planet. Overnight gaps are concentrated around the open, when price discovery is loud and the book is re-forming. A stop that triggers into the open can get a fill that has almost nothing to do with "value." If you insist on using stop orders, consider whether you want them active overnight at all. For some investors, the more rational choice is to manage exit decisions during liquid hours, and rely on sizing and hedges for the discontinuous risk that lives outside the session.

None of this is meant to make markets sound villainous. Markets are doing an honest job of repricing uncertainty. The villain is the fantasy of continuous, friendly liquidity—the idea that you can always leave at a price that respects your plan. In calm conditions, that fantasy often survives. In crisis conditions, it doesn't. The investors who come through crises intact are not necessarily the ones with the cleverest forecasts. They're the ones whose protection doesn't rely on the market granting them a favor at the worst moment.

Actionable takeaway: Pick one core position in your portfolio and write down the answer to a blunt question: *"What is my maximum tolerable overnight gap loss in dollars?"* If your honest answer is "I need a floor," don't outsource that need to a stop-loss. Price out a defined-risk hedge (for example, a put at a strike that matches your floor) and compare its premium to the damage a gap would do. If you decide the premium is too high, that's useful information: it's the market telling you the risk is real. The next move isn't to hope—it's to reduce exposure to a size you can actually carry through a discontinuity.

6.3 The Margin Call Nightmare

The most painful part of a margin call is not the phone notification or the stern email. It's the timing. Margin calls arrive when your portfolio is already bleeding, when headlines are screaming, and when every sensible cell in your body is begging you to do *nothing* rash. Then your broker effectively says: *Move. Now. Or we will move for you.*

That's the nightmare scenario for a long-term investor using hedges: you can be directionally right about the market, you can even have a hedge that's gaining value, and still get forced to liquidate your best long-term holdings at the worst possible moment because you ran out of ready cash.

Margin is borrowing, but it behaves like a landlord. It's friendly on move-in day, strict when the neighborhood gets rough, and it doesn't care about your investment horizon. In practical terms, margin is a rule set that determines how much equity you must maintain in your account. If losses (or changing requirements) push you below that line, you have a deficit. The deficit is not a debate. It's a deadline.

Most people think margin calls happen because you did something "aggressive," like buying meme stocks with borrowed money. Sometimes, yes. But the more subtle way margin bites hedgers is through *collateral mechanics*—the behind-the-scenes cash needed to support positions that can lose money fast, or that your broker treats as potentially dangerous in a stressed market.

Here's the core misunderstanding that causes intelligent investors to get blindsided: your account can be safe in the long run and still be unacceptable to a margin system in the short run. Brokers and clearinghouses are not trying to predict what your portfolio will do over the next five years. They're trying to make sure they get paid if your positions move against you *today* and you can't.

A margin account has a few moving parts, but you don't need a finance degree to understand the failure mode. Your account equity is basically: *what you own* minus *what you owe*. Your broker sets maintenance requirements—a minimum equity level. Fall below it, and you either deposit money/securities or positions get liquidated. Not "might." *Can.* And often the customer agreement gives the broker wide discretion on timing, amount, and which positions get sold.

That last detail is where the nightmare becomes a horror movie. When you're in deficit, you don't get to say, "Sell this loser first." Many margin agreements allow the broker to liquidate without contacting you and to choose what gets sold to fix the shortfall. If you pictured margin calls as a polite negotiation, replace that mental image with a fire department breaking down a door. They're not asking permission; they're preventing a bigger problem.

Hedging strategies can create margin risk in places that don't feel like "leverage." Consider selling options to finance a hedge—a collar, for example, where you buy a put and sell a call. The purchased put is defined-risk; the sold call is not (at least not in the same way). If the market whipsaws upward, the short call can generate losses and margin requirements right when you least expected it. Or consider spreads that look limited-risk on paper but still require margin because the broker needs collateral against assignment risk, early exercise risk, or gap risk.

Even if you never sell an option, you can run into margin trouble if you hedge with instruments that settle daily, or that can move violently relative to posted collateral. In the cleared derivatives world, two terms matter: *initial margin* and *variation margin*. Initial margin is the "good faith" deposit meant to cover potential losses under stressed scenarios. Variation margin is the daily (sometimes intraday) transfer of profits and losses. If your position loses value, cash moves out of your account. If it gains, cash moves in. This process keeps the system stable—but for the individual investor, it can turn volatility into immediate cash needs.

That cash timing is the villain. Long-term investors can tolerate volatility in net worth. They cannot tolerate a demand for cash on a short fuse if their cash is tied up in investments they don't want to sell. A margin call converts volatility into a liquidity event. And once it's a liquidity event, you are no longer making decisions based on strategy. You're making decisions based on what can be sold fast enough.

This is where the execution reality from earlier becomes relevant without needing to repeat it: when markets are stressed, selling quickly tends to be expensive. If you are forced to liquidate, you are often forced to liquidate into wider spreads and thinner depth. A margin call doesn't just force you to sell; it forces you to accept the market's least generous terms.

The most common path into a margin call is painfully ordinary: you hold stocks or ETFs, maybe a hedge overlay, maybe a little leverage, and then volatility spikes. In response, requirements change. Brokers can have "house" margin requirements that are stricter than baseline regulatory minimums, and they can raise them when conditions deteriorate. This is not personal. It's risk management. But from your perspective, it means your account can become "more leveraged" overnight without you trading a single share—because the definition of required collateral just moved.

Now add the second accelerant: concentration. A concentrated position can become "special" in the broker's risk system very quickly. A stock that was treated as plain vanilla yesterday can be reclassified as higher risk today. Requirements can rise. If you're using that position as the backbone of your long-term plan, a sudden hike in maintenance can force partial liquidation at a terrible time, turning a temporary drawdown into a permanent loss of shares you intended to hold for years.

Assignment risk deserves special attention because it's the margin call nightmare in costume. If you sold an option and it gets exercised, you can wake up owning (or short) a pile of stock you never intended to hold. That new stock position has its own margin requirement. The exercise itself can create a cash debit. And the timing can be absurd: options can be exercised early, and the market rarely asks whether you had a quiet week planned. Assignment isn't inherently bad, but it's operationally demanding. The investors who get hurt aren't the ones who don't understand options theory; they're the ones who didn't plan for the cash-and-collateral mechanics of what happens when theory becomes a real position.

There's also a psychological cruelty to margin: it makes you feel like your own hedge betrayed you. Imagine you put on a hedge that costs premium, the market drops, and the hedge starts working. Meanwhile, your long positions are down, your broker tightens requirements, and you get a margin call. You sell your longs near the lows to raise cash. Then the market rebounds, and your hedge loses value as fear fades. You end up with fewer long-term shares *and* an insurance premium you paid for protection that arrived too late to prevent the forced sale. The hedge didn't fail. Your liquidity plan did.

The fix is not glamorous. It's cash management, and it's unreasonably effective.

Start by treating margin like a stress test you must pass to be allowed to keep your positions. The question isn't "How much leverage can I get?" The question is "How much drawdown and how much requirement-tightening can I survive *without* being forced to sell?" That's a different mindset. It turns margin capacity from a spending limit into a safety buffer.

A good operational habit is to maintain a dedicated *cash buffer*—money or highly liquid short-duration instruments that you mentally label as "margin oxygen." Not emergency savings for your life. Emergency oxygen for your account. The point is to have deployable collateral that does not require selling depressed assets. The most dangerous portfolio is the one that is 100% invested and 0% liquid while using instruments that can generate sudden cash demands.

How large should the buffer be? There's no universal number, but you can build a sane range with a simple thought experiment. Imagine an ugly week where your equity holdings drop 10% and your broker raises requirements at the same time. Imagine you have an options overlay and implied volatility spikes, increasing the margin needed for any short option exposure. Ask: *What dollar amount of cash could I be asked for on short notice?* Then assume your first estimate is optimistic, and add a margin of safety to your margin safety.

If you use hedges that can require variation margin (or you're trading in a way that creates daily cash flows), the buffer should be more conservative. Variation margin is like a daily audit of your mark-to-market reality. It's fair. It's also relentless. If your thesis is right but your timing is wrong, you can still be drained by daily losses before you get the payoff you expected.

The next habit is to separate "collateral assets" from "core holdings." Many investors let the broker's default behavior decide what gets liquidated. That's how you lose the crown jewels. Instead, consider keeping core, long-term holdings in a structure less exposed to forced liquidation risk, and using a separate, deliberately funded account (or sub-portfolio) for strategies that can generate margin calls. The operational goal is simple: if the hedge machine breaks, it should not eat the retirement portfolio.

That separation is not about hiding from risk; it's about controlling blast radius. A hedge is supposed to reduce downside, not create a new path to ruin that bypasses market returns entirely.

It also helps to treat option selling with the seriousness it deserves. Selling options can be a sensible part of hedging (financing puts, generating income, shaping payoffs), but it's not free money. When you sell an option, you are selling insurance to someone else. Insurance companies don't run fully invested with no liquidity. They maintain reserves. If you want the benefits of premium collection, you need the temperament and the balance sheet behavior of an insurer—including the willingness to hold cash that feels "unproductive" most of the time.

A common retail mistake is to use short options as a casual coupon: "I'll sell a put to get paid while I wait." Then volatility spikes, the short put explodes in value, margin requirements rise, and the investor is suddenly managing a position that behaves like a leveraged bet. If you are going to sell puts, do it with a pre-committed plan for what happens if you get assigned, and with cash earmarked as if assignment is not a remote possibility but a scenario on the menu.

Another operational trick: be careful with "all-weather" language. Investors say, "This is a conservative strategy," when they mean, "In normal conditions, it behaves calmly." Margin systems don't care about normal. They care about stress. If a position can blow out under stress, it is treated as stress-prone.

The market's job is to surprise you; your job is to not be fragile when it does.

You can also reduce margin fragility by choosing hedges that are *fully paid*. Buying puts costs premium, but the maximum loss is known upfront. That can be psychologically and operationally stabilizing. Defined-risk spreads can also be useful, though you should learn exactly how your broker margins them and what assignment could do to the account overnight. The point isn't "never use margin." The point is to understand whether your hedge relies on borrowed time and borrowed money.

One more underappreciated risk is correlation. Investors plan for "stocks down" and "hedge up." In real stress, correlations can behave badly. Assets you thought were diversifiers can sell off together because investors are raising cash. If your "safe bucket" declines while your margin needs rise, your buffer shrinks precisely when you need it most. This is why cash is boring but powerful: cash doesn't correlate. It just sits there, waiting to solve problems.

Finally, there's the operational readiness piece. A margin call is a time-pressure event, so you want a time-pressure plan. If you had to raise $10,000 by tomorrow morning, do you know which assets you would sell? Do you know which trades you would close first? Do you know what you refuse to sell? In a crisis, thinking is expensive. Pre-deciding is cheap.

The margin call nightmare ends the same way it begins: with a deadline. The investor who survives is the one who can meet that deadline without torching the long-term plan. That's not about cleverness. It's about liquidity, redundancy, and humility.

Actionable takeaway: Write a one-page *margin fire drill* for your own account. Include:

- your current cash buffer amount,
- the positions that could trigger sudden margin needs (any leverage, any short options, any products with daily settlement),
- a ranked list of what you would liquidate first to raise cash,
- a minimum "do-not-cross" cash buffer you will maintain going forward.

Then, the next time you add a new hedge or strategy, update the fire drill *before* you place the trade—if you can't describe how it behaves under stress, you don't own a hedge, you own a surprise.

6.4 Is Your Insurance Solvent? Counterparty Risk

A hedge is supposed to be the thing that still works when everything else is misbehaving. That's the whole emotional bargain: when your portfolio is down and your stomach is tight, the hedge should be the adult in the room. But there's an awkward question most investors don't ask until they're forced to: *What if the hedge depends on someone else staying alive?*

That's counterparty risk in plain English. It's the risk that the entity on the other side of your trade—or the institution standing between you and your assets—can't perform when it matters. In good times, it feels like a pedantic footnote. In bad times, it can be the difference between "my hedge didn't pay off" and "my hedge doesn't even exist in the way I thought."

Counterparty risk is the perfect partner to the execution problems you've already seen. Liquidity risk is the market saying, "I can't give you a good price right now." Counterparty risk is the market saying, "I might not be able to deliver the thing you believe you own." One is about *price*. The other is about *structure*.

The simplest way to feel the difference is to imagine buying insurance from two places. One is a regulated insurer that holds reserves and is monitored. The other is a stranger who says, "Trust me, I'll pay you if your house burns down," and hands you a business card. Both might be cheap. Only one is a real plan.

For retail investors, the most common structural trap is hiding in plain sight: a ticker symbol that looks like an ETF but is actually an ETN.

An *ETF* (Exchange-Traded Fund) is typically a fund that holds assets (stocks, bonds, futures, swaps, cash) inside a legal structure, often a separate pool or trust. When you buy shares, you own a slice of that pool. You have market risk—if the assets drop, you drop—but you generally are not relying on a bank's promise in the same direct way. The ETF sponsor can be replaced; the assets are meant to remain in the fund.

An *ETN* (Exchange-Traded Note), by contrast, is usually unsecured debt issued by a bank. It's an IOU that says, "We promise to pay you a return linked to some index." The note trades on an exchange like a stock, so it *feels* like a fund. But structurally, you're a creditor of the issuer. If the issuer gets into trouble, your "hedge" becomes a claim in a bankruptcy queue. In other words: you didn't buy a basket of assets. You bought a promise.

When people hear this, they often react with disbelief because it clashes with the user experience of trading. The symbol looks clean. The bid-ask spread looks tight. The chart looks normal. But the label on the bottle matters. A fund is one thing. A note is another. The market can make two different products look identical on a brokerage app, the way two bottles can look identical on a shelf. Then one day you learn one bottle was water and the other was perfume.

Why would anyone buy an ETN? Because ETNs can offer exposure that's hard to package in an ETF, and they can track an index in a very direct way. For certain exotic exposures, an ETN can be a convenient wrapper. Convenience is not the enemy. Unexamined convenience is.

The key mental shift is to stop thinking in tickers and start thinking in *balance sheets*. If your hedge is an ETN, the relevant question isn't only "Will this index go up when I need it?" It's "Will this issuer be solvent and operationally functional when everyone is panicking?" That's a very different bet.

A useful historical anchor is what happened in the 2008 era, when household-name financial institutions went from "too big to worry about" to "gone" with shocking speed. That period burned a simple truth into market memory: the worst time to discover you're holding someone else's unsecured promise is when credit spreads are exploding.

Even if your hedge is not an ETN, counterparty risk can still sneak in through the plumbing of the brokerage system. Many investors use "counterparty risk" as shorthand for "the person on the other side of my trade," but for retail investors the more immediate question is often *custody*: if your brokerage firm fails, what happens to the stocks, ETFs, and cash you believe are sitting in your account?

In the U.S., most mainstream broker-dealers are members of SIPC (the Securities Investor Protection Corporation). SIPC protection is widely misunderstood because the name sounds like insurance against loss. It is not. SIPC is primarily about returning customer securities and cash that were supposed to be in your account if the brokerage fails and there's a shortfall. It's about custody and recordkeeping failures, not about the market going down, not about your hedge not working, and not about bad timing.

That distinction matters because it changes the way you evaluate "safety." If you buy a put option and the market crashes, SIPC is irrelevant (you're just experiencing market moves). If your broker fails and your securities are missing or tied up, SIPC may matter a lot—but it still won't make a bad trade good. It will try to make you whole on what you *already owned* at that brokerage, within its rules and limits.

Zoom out one layer and you run into the architecture that makes listed derivatives feel boring (in the best way): central clearing.

When you trade listed options in the U.S., you're not relying on the random person who sold you the option to pay up. The system uses a clearinghouse model. The Options Clearing Corporation (OCC) stands in the middle, becoming the buyer to every seller and the seller to every buyer. This is one of the reasons listed options are such a powerful hedging tool: they convert messy bilateral counterparty exposure into a standardized, margined, centrally cleared system.

But central clearing doesn't eliminate risk. It reshapes it. It reduces the risk that your specific counterparty fails, while concentrating systemic responsibilities in the clearinghouse and its members. Clearinghouses manage this with layers of protection often described as a "default waterfall": margin posted by members, guarantee funds, and additional assessment powers under certain conditions. You don't need to memorize the waterfall to benefit from it, but you should understand what it implies: listed markets are engineered to survive member defaults, *at the cost* of forcing participants to post margin and meet cash calls promptly.

That connects directly to the margin-call reality you've already felt: the system's safety is purchased with collateral discipline. When volatility spikes, margin requirements can rise. When losses occur, variation margin moves cash quickly. This is how a system avoids the slow-motion contagion that can occur when losses are allowed to linger as unsecured promises.

Futures markets add another important structural concept: segregation of customer funds. Futures commission merchants (FCMs) are generally required to segregate customer margin funds from the firm's own money, with specific legal protections around how those funds are held. Again, this doesn't protect you from the market moving against you. It's not a profit guarantee. It's an attempt to reduce the damage if the intermediary fails.

So where does this leave the long-term investor who just wants downside protection without a PhD in market structure?

It leaves you with a surprisingly practical checklist: *What am I actually holding, and what failure mode am I assuming away?*

If you're using an ETF for hedging exposure—say, a broad index ETF, a bond ETF, or a volatility-related ETF—you're taking market risk in the assets and strategy the fund runs. You still need to care about liquidity, tracking, and how the fund gets exposure. But you're generally not making a direct credit bet on an issuer's solvency in the same way you are with an ETN.

If you're using an ETN, you are explicitly accepting issuer credit risk. That may be a reasonable trade in some contexts, but it's not "just another ETF." It is a different animal with a different set of questions: Who issues it? What is the note's structure? How does it behave if the issuer is downgraded, stressed, or fails? Are there call features or other terms that can change your outcome at the wrong time? (Many notes have provisions that can surprise investors who assumed the product was a simple perpetual fund.)

There's also a softer form of counterparty risk that doesn't show up as default, but as *product behavior under stress*. Some instruments can be halted, delisted, redeemed, or otherwise changed by rules and prospectus mechanics. This is not a conspiracy; it's the legal structure asserting itself. In a crisis, legal structure matters more than marketing. Your hedge plan should be compatible with the product's "break glass" clauses.

This is why it's dangerous to treat all "exchange-traded" things as morally equivalent. Exchange-traded tells you where it trades, not what it is. A tuna sandwich and a peanut butter sandwich can both be "sandwich." That label is not enough if you're allergic.

A common mistake is to confuse *complexity* with *danger*. Some simple products carry meaningful counterparty exposure (an ETN can be simple to understand and still be a credit bet). Some complex products are built on robust infrastructure (listed options can be complex in payoff but centrally cleared). The right way to judge is not by how intimidating the payoff diagram looks. Judge by who owes what to whom if the world gets ugly.

There's also a temptation to soothe yourself with the thought, "If something that big fails, we'll have bigger problems." That's emotionally comforting and practically useless. Your job isn't to predict whether the financial system survives. Your job is to avoid unnecessary fragility in your own plan. If you can choose a hedge instrument that doesn't require you to be right about a specific institution's creditworthiness, why wouldn't you?

The good news is that you don't need to become paranoid to be prudent. Counterparty awareness is not the same thing as counterparty fear. It's simply refusing to outsource your safety to a label.

A realistic, retail-friendly framework is to separate counterparty risk into three buckets:

- *Product structure risk:* Is the instrument a fund holding assets (ETF) or a note that is an unsecured promise (ETN)? If it's a derivative exposure, is it listed and cleared, or is it over-the-counter in a way you can't easily evaluate?
- *Intermediary/custody risk:* What happens if your broker fails? Are you at a SIPC-member broker? Do you understand that SIPC is about returning custody assets, not covering market losses?
- *What is explicitly not insured:* No matter how "protected" you feel, market losses are still yours. A hedge can reduce them, but it doesn't outsource them to SIPC, to a clearinghouse, or to anyone else. Those systems are about settlement integrity, not investment outcomes.

That last bucket sounds obvious, but it saves people from a specific kind of disappointment: the belief that "safety" means "no losses." Safety in hedging is narrower and more honest. It means the contracts are likely to settle as designed, and the structure is less likely to fail at the exact moment your emotions are most raw.

If you want a single, vivid test to apply before buying any hedge product, use this question: *If this hedge works perfectly economically, could it still fail operationally?* If the answer is yes, identify the "who" behind the failure. Is it an issuer? A broker? A clearing system? Then decide whether you're being paid to take that risk. If you're not being paid, don't take it.

Actionable takeaway: Before your next hedge trade, pull up the product's issuer/structure description and write one sentence beginning with: *"I am relying on ___ to pay/hold/settle this."* If you can't fill in the blank clearly, pause. If the blank is "a bank's promise" (ETN), decide explicitly whether you are comfortable taking issuer credit risk for your hedge. If you'd rather not, choose a structurally different instrument (often an ETF or listed options) where your downside plan depends more on market mechanics and less on a single institution's solvency.

6.5 Crash Test Dummies: Stress Testing

The day you *need* a hedge is the day your brain becomes a novelist. It can write a convincing story in seconds: "My puts will cover this," or "My bond sleeve will hold up," or the classic, "I'll just sell if it gets worse." In calm markets those stories feel harmless. In real drawdowns they become expensive fiction.

Stress testing is how you replace fiction with numbers before the market starts charging tuition.

A good stress test doesn't try to predict the next crisis. It tries to answer a simpler question: *If the world turns ugly in a familiar way, what happens to me?* Not to the S&P 500. Not to a theoretical portfolio. To your actual positions, with your actual sizes, your actual liquidity constraints, and your actual habits under pressure.

The best part is that you don't need institutional software to do this well. You need a spreadsheet, a willingness to be slightly paranoid, and the humility to assume that in a crisis you will not execute at the midpoint, you will not get a perfect fill, and you will not feel clever.

There are two kinds of stress tests that matter for hedging. The first is *P&L stress*: how much you lose if markets drop. The second is *operational stress*: whether you can survive the cash, collateral, and execution realities that show up at the same time. Most investors do the first one (if they do any at all). The second one is where the bodies are buried.

Start with the P&L stress because it's the skeleton that everything else hangs on. Pick a scenario that makes your palms sweat but doesn't feel like science fiction. "S&P down 20%" is a perfectly serviceable baseline, because it's large enough to matter and common enough to be plausible. You're not forecasting; you're measuring sensitivity.

Then translate that scenario into *your* portfolio's language. If you own a diversified basket of U.S. equities and equity ETFs, your first approximation is to assume those holdings drop roughly with the market. If you own higher-beta growth stocks, assume they drop more. If you own defensive stocks, maybe less. Precision is nice, but honesty is better: your goal is to avoid the surprise that arrives when you discover your portfolio was more fragile than you believed.

Now bring the hedge into the same scenario. If you own protective puts, estimate their value increase under the drop. If you own a collar, remember the short call may soften gains if the market whipsaws upward later, but in the down leg the put is doing the heavy lifting. If you hedge with something like an inverse ETF, model what it does in a down move *and* remember that path matters (compounding can make inverse products behave differently than you expect over time). Keep the math simple: a stress test that you can run every quarter beats a perfect model you run once.

Here's a concrete example to make it real. Imagine a $500,000 portfolio that's roughly 80% equities and 20% cash/bonds. In a 20% equity drawdown, the equity sleeve loses about $80,000 (20% of $400,000). If you bought enough puts to cover half the equity exposure for that move, maybe the puts gain $35,000—$45,000 depending on strikes, time to expiration, and volatility. Suddenly your loss is not "$80,000" but "$35,000—$45,000 plus premiums already paid." That difference is not academic. It changes your behavior. It can be the difference between panicking into a permanent mistake and doing what long-term investing requires: continuing to own assets through discomfort.

Stress testing becomes truly useful when you stop treating it like a single number and start treating it like a range. In a crisis, correlations tighten, volatility jumps, and the market develops sharp elbows. Your hedge may perform better than expected because volatility explodes and puts become more valuable. Or it may perform worse if you were under-hedged, if your hedge is illiquid, or if your instrument doesn't behave cleanly under stress. Ranges force you to admit uncertainty *in advance,* which is the only time uncertainty is cheap.

Now add the operational layer: the crisis doesn't just change prices; it changes *requirements*. Margin can increase. Bid-ask spreads widen. Depth thins. If your hedge involves any position that can create cash demands (short options, leveraged products, futures, or even a broker raising house requirements), your stress test should include a cash call assumption. Think of it as a second axis: not "How much do I lose?" but "How much cash might I need *while* I'm losing?"

This is where many stress tests fail in the most human way possible: they assume you can sit still. Real crises often don't let you sit still. Your broker may not care that your long-term plan is sound. Your clearing system may not care that your hedge will eventually pay. When collateral rules say "pay now," your future gains don't count as cash in the present.

A good operational stress test therefore asks an unromantic question: *Could I meet a sudden demand for $X in cash without selling my long-term holdings into a wide-spread market?* If the answer is no, the portfolio is not merely risky; it is fragile. Fragile portfolios don't just lose money. They lose *optionality*—the ability to choose what to do next.

If this is starting to sound like professional risk management, that's because it is. The good news is the retail version is simpler, not weaker. Professionals often run dozens of scenarios because they manage huge books and complex constraints. You can get most of the benefit by running a handful of scenarios that map to your actual fears: a sharp equity drop, a volatility spike, and a liquidity squeeze that forces ugly execution.

Historical scenarios help because they come with emotional texture. 2008 teaches you about long, grinding drawdowns and the feeling of helplessness. 2020 teaches you about speed—how quickly markets can fall and how quickly they can bounce, punishing people who assumed they'd have time to "get out." You don't need the exact returns of those years to learn from them. You need the shape of the pain.

A particularly useful stress test is the "two-hit combo." First, assume equities drop 20%. Second, assume volatility jumps enough that option spreads widen and the cost to roll hedges increases. Third, assume your broker tightens margin requirements if you have any positions that could be margin-sensitive. The numbers don't have to be perfect. The logic does. Crises rarely arrive as a single clean punch; they arrive as a pile-on.

This is also the moment to stop grading your hedge on a single headline outcome like "Did I make money?" A hedge can be a success even if it loses money, if it prevented you from making a catastrophic behavioral decision. Likewise, a hedge can be a failure even if it makes money, if it introduced operational fragility that forced you to liquidate your core holdings.

One of the most useful outputs of stress testing is not the loss estimate. It's the hedge sizing insight. When you stress test, you discover whether you bought *comfort* or *coverage*. Comfort hedges are small enough to feel responsible but too small to matter when the market drops. Coverage hedges meaningfully change the drawdown profile. The difference is not philosophical; it's arithmetic.

The second output is humility about execution. If your stress test assumes you can "rebalance into the dip," remember that rebalancing requires trading. Trading requires liquidity. Liquidity is least friendly when everyone else wants to do the same thing. A stress test that includes an execution penalty—even a simple one like "assume I lose an extra 0.5% to 1.5% in slippage and spreads when I trade under stress"—is often more realistic than a pristine model that assumes you always get the mid.

To make the point concrete, here's a simple visual that compares how an unhedged portfolio can behave versus a hedged one in two widely remembered crisis episodes. The specific numbers are illustrative, but the shape is the message: hedging is not about eliminating loss; it's about shrinking the hole to a size you can live with.

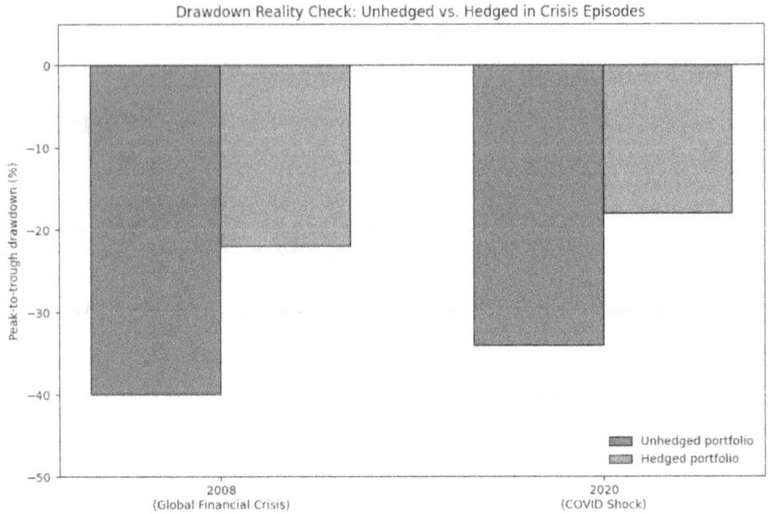

The picture isn't telling you to fear -40%. It's telling you to be honest about what -40% does to a human being with a login button. A hedged drawdown that's "only" -18% or -22% can be the difference between staying invested and detonating your plan at the bottom. The math changes the psychology, and the psychology changes the outcome.

A final, underappreciated benefit of stress testing is that it forces you to define success *before* the crisis. If you wait until things are falling, you'll move the goalposts. You'll decide the hedge was "too expensive" right after it saves you, or "too small" right after it fails to prevent pain. If you define success ahead of time—for example, "I want to cap my portfolio drawdown at roughly 20% in a 30% equity drawdown"—you can evaluate the hedge on its job description, not on your mood.

Stress testing also helps you avoid the most common hedging regret: paying premiums for years and then abandoning the hedge right before it would have mattered. That regret often comes from buying a hedge without understanding what it's supposed to do in a bad scenario. When you stress test, you see the trade clearly: you're paying small, frequent costs to reduce the probability of a large, life-altering mistake. That trade feels different when you've run the numbers.

The goal isn't to build a bunker. It's to build a portfolio you can actually hold through weather. That means knowing what happens when markets gap, when liquidity dries up, when margin rules tighten, and when your own patience gets tested. Stress testing is the rehearsal that makes the performance less chaotic.

Actionable takeaway: Run a quarterly "two-page" stress test for your real portfolio. Page one: assume the S&P 500 drops 20% tomorrow and estimate your portfolio loss using simple betas or rough exposure weights; estimate your hedge gain using conservative assumptions (don't use the best-case). Page two: assume spreads widen and margin/collateral needs rise—write down how much cash you could access within 24 hours without selling core holdings. If the cash number is uncomfortably small, the fix is not more optimism; it's a bigger cash buffer, a less margin-sensitive hedge, or smaller risk in positions that can force your hand.

6.6 The Scorecard: Metrics for Success

There's a special kind of regret that only hedgers experience: paying for protection for months, then watching the market drift upward and thinking, *I wasted money*. And then, in the next drawdown, ripping the hedge off in frustration—right before it would have mattered. This isn't a character flaw. It's what happens when you try to judge a hedge the way you judge a stock pick: by short-term profit.

A hedge needs a different scoreboard. It's not trying to "beat the market." It's trying to change the shape of your outcomes: less gut-wrenching drawdowns, fewer forced decisions, more ability to stick with the plan. If you don't measure those things, you end up measuring the one thing a hedge is almost designed to fail at during calm markets: making you feel smart.

The simplest way to stay honest is to adopt a few risk metrics—not as badges of sophistication, but as instruments on a dashboard. Pilots don't stare at the scenery and guess their altitude. Investors shouldn't stare at a portfolio and guess their risk.

Three metrics do most of the work for long-term investors: *beta* (how much you tend to move with the market), *correlation* (whether your hedge really zigzags when your portfolio zags), and *Value at Risk* or *VaR* (a disciplined way to translate volatility into "how bad could it get" over a short window). None of these are perfect. All of them are better than vibes.

Beta: the volume knob on market risk

Beta is the least intimidating metric in finance once you stop treating it like a secret. Beta is simply how sensitive your portfolio is to a benchmark, usually the S&P 500. If the market moves 1% and your portfolio tends to move 1%, your beta is about 1. If the market moves 1% and you tend to move 0.6%, your beta is about 0.6.

The most useful way to think about beta is as a volume knob. A beta of 1 turns the market's moves up to full volume. A beta of 0.5 turns them down. A beta above 1 is like leaning closer to the speaker. Hedging, at its core, is an attempt to turn the volume down without turning the music off.

A common hedging failure is "expensive silence": you pay option premium, but your portfolio still behaves like a beta-1.0 animal because the hedge is too small, too short-dated, or too far out-of-the-money to matter until the crash is already severe. The hedge is not useless—it may help in the tail—but it is not doing the everyday job many investors subconsciously want: reducing the portfolio's day-to-day market sensitivity.

Beta gives you an immediate gut-check. If you believe you've meaningfully hedged equity risk, your realized beta should generally fall. If it doesn't, you didn't buy a hedge; you bought a lottery ticket that only pays in a specific kind of disaster.

There's a catch, and it's an honest one: beta is not a constant. In calm markets, diversified stock portfolios often behave with stable beta. In crises, correlations jump and betas can drift higher just when you wish they wouldn't. A hedge that looks "small" in beta terms during quiet months can suddenly become powerful when volatility spikes (because options are nonlinear). So beta isn't a verdict. It's a reading on the dashboard that tells you whether the hedge is influencing your ride most of the time.

If you want a practical way to use beta without building a quant lab, do this: pick a lookback window (say, the last 6 or 12 months of weekly returns) and estimate your portfolio's beta to the S&P 500. Many broker platforms and portfolio tools will do this for you; you can also do it in a spreadsheet using a simple regression. Then track it over time. You're not hunting for perfection. You're watching for drift.

A useful gut check: if you're trying to run a "sleep at night" portfolio, you might aim for a beta that's meaningfully below 1—not necessarily tiny, just intentionally lower. The exact target depends on your goals, age, and cash flow, but the key is that you should be able to articulate a target and notice when you've wandered away from it.

Correlation: the lie detector for diversification

Correlation is the answer to a deceptively simple question: *When my portfolio is down, is this hedge typically up?* If the answer is "usually," correlation is negative and the hedge is doing its diversifying job. If the answer is "sometimes, but not when it counts," correlation may be close to zero in calm periods and shoot toward one in stress—the classic heartbreak of "diversifiers" that suddenly behave like stocks when the room catches fire.

Correlation is the lie detector because marketing language loves the word "uncorrelated." In practice, many assets are uncorrelated right up until the moment investors need cash, and then they sell everything they can. Correlations are not laws of nature. They're crowd behavior with math on top.

When you evaluate a hedge, correlation keeps you from being seduced by a good story. Maybe you bought something because it sounded like protection—a "defensive" sector, a factor tilt, a foreign market, a strategy ETF with clever language. Correlation asks the blunt question: *Did it actually behave differently when your portfolio suffered?*

The trick is to measure correlation in at least two regimes: normal times and ugly times. If you only look at full-sample correlation, you can miss the most important feature: how the relationship changes during drawdowns. A hedge that is mildly helpful in normal times but strongly helpful during selloffs can be worth real money. A "diversifier" that looks uncorrelated in normal times but becomes highly correlated in selloffs can be worse than useless—it can give you false confidence and encourage you to take more equity risk than you should.

A retail-friendly way to do this is to compute correlation using weekly returns over the last year, then compute it again using only the weeks when the market was down more than, say, 2%. You're not building a thesis for publication. You're trying to see whether your hedge tends to show up when you call its name.

VaR: translating volatility into a worst-case headline

VaR, or Value at Risk, has an intimidating reputation because it's often presented with Greek letters and smugness. Strip that away and VaR is simply a disciplined way to say: "*On most days, I don't expect to lose more than $X.*" Usually it's stated with a confidence level and a time horizon, like "one-day 95% VaR." That means: based on recent volatility and assumptions, there's a 95% chance you won't lose more than $X in a day. The other 5% is the scary tail where bigger losses live.

VaR is useful for hedgers because it forces you to speak the language your nervous system understands: dollars. A 1.2% daily volatility number doesn't punch you in the gut. A "$12,000 expected worst day (95%)" number does.

Here's the important honesty clause: VaR is not a crash predictor and it is not a guarantee. In real crises, losses can exceed VaR dramatically. VaR is a seatbelt, not an airbag. But as a monitoring tool, it's valuable because it moves with your portfolio's risk. If your hedge is doing its job, your VaR should generally fall, or at least rise less during turbulent periods than it would without the hedge.

A practical VaR for retail investors can be as simple as this: compute the standard deviation of your portfolio's daily (or weekly) returns over a lookback period, multiply by a z-score (about 1.65 for 95%, about 2.33 for 99%), and translate it into dollars by multiplying by portfolio value. This is a simplified "parametric" VaR. It's not perfect, but it's consistent, and consistency is what makes a dashboard useful.

If you prefer to avoid assumptions about normal distributions, you can do historical VaR: look at your worst daily (or weekly) returns in the sample and pick the percentile. Again, imperfect, but emotionally honest.

What VaR adds to beta and correlation is a sense of absolute scale. Beta can tell you you're less sensitive than the market. Correlation can tell you your hedge is behaving differently. VaR tells you whether the remaining risk is still too large for your real life.

Measuring the hidden leak: execution and ``implementation shortfall"

Risk metrics can tell you whether the *idea* of your hedge is working. They don't automatically tell you whether your *trades* were executed efficiently.

This is where a simple execution metric helps: *implementation shortfall*. It's a formal term for an everyday experience: the difference between the price you *intended* to get and the price you *actually* got, including the opportunity cost of not getting filled when you wanted.

You don't need to calculate it with institutional precision to get value from it. The retail version is a trading journal with two numbers for each hedge trade: the midpoint when you decided to trade and the actual fill price. Record the spread at the time. Over a few months, patterns emerge. You'll see which instruments quietly tax you, which order types hurt you, and whether your "cheap hedge" is being eaten alive by wide spreads and slippage during stressed periods.

This matters because hedging can fail through a thousand paper cuts. You can design a beautiful hedge and then donate its edge to the market by trading it poorly—chasing volatility, using market orders when liquidity is thin, rolling at the worst times. A scorecard that ignores execution is like a diet plan that ignores snacking.

A workable scorecard for real humans

The best scorecard is the one you'll actually use, which means it must be short, repeatable, and tied to decisions. Consider tracking these items monthly or quarterly:

- *Portfolio beta* to your chosen benchmark (e.g., S&P 500) and whether it's near your intended range.
- *Hedge correlation* with your equity sleeve, measured in normal weeks and down-market weeks.
- *Simple VaR* (in dollars) so you can answer, "What's a bad week likely to cost me?"
- *Hedge carry cost* (premiums paid minus hedge payouts realized) so you stay honest about insurance expense.
- *Execution leak* (your rough implementation shortfall) so you can see whether trading friction is quietly sabotaging you.

That may look like a lot, but in practice it's a single page. More importantly, it stops you from doing the most common destructive thing hedgers do: changing the hedge because they're bored, embarrassed, or anchoring to the last market mood.

A hedge will almost always look dumb right before it looks brilliant. Metrics don't prevent that emotional whiplash, but they give you a stabilizing reference point. They allow you to say, with a straight face, "The hedge is doing what it was hired to do," even when the market is trying to bait you into impulsive changes.

The final step is to connect metrics to action. A scorecard without consequences becomes decoration. If your beta has crept up, either increase hedging, reduce equity exposure, or admit you're comfortable with more risk than you previously claimed. If correlation has become less negative in down markets, reconsider whether the hedge is the right instrument. If VaR is uncomfortably high, your "long-term" horizon may be irrelevant because the short-term swings are large enough to force behavioral or margin-driven decisions.

None of these metrics is a moral judgment. They are just a mirror. The goal is not to look heroic in the mirror. The goal is to avoid being surprised by your own reflection when markets get ugly.

Actionable takeaway: Create a one-page hedging dashboard and update it on a fixed schedule (monthly or quarterly). Include: your portfolio's beta to the S&P 500, the correlation of your hedge to your equity sleeve during down-market weeks, and a simple VaR estimate in dollars. Add a small execution journal line item: record the mid-price and your actual fill for each hedge trade. If the metrics say you're taking more risk than intended (beta/VaR rising) or your hedge isn't behaving as protection (correlation drifting toward positive in down weeks), make one concrete adjustment and write down why—so the next change is driven by evidence, not emotion.

Chapter 7

Contexts, Taxes, and Special Cases

It is not just about what you buy; it is about where you hold it and what you keep after the IRS takes its cut. In the real world, even a brilliant hedging strategy can fail if taxes eat up your profits or if you attempt to execute it in an account that strictly forbids it. This chapter moves from the whiteboard to the messy reality of your financial life, helping you navigate complex tax rules, rigid retirement account constraints, and the high-stakes challenge of protecting a concentrated nest egg without triggering a massive tax bill.

7.1 The Taxman Cometh: Implications for Hedgers

Picture a homeowner who installs a top-of-the-line security system. Cameras, sensors, the whole thing. Then they discover the "monthly monitoring fee" is only part of the cost: the system also trips the circuit breaker often enough that they keep restocking the fridge after spoiled-food incidents. That's the tax version of hedging. The premium you pay for protection is obvious; the *tax drag*—the quiet leakage caused by how hedges are taxed—is the cost people notice only after the fact, usually while staring at a surprisingly large April bill.

Hedging tends to create more transactions, and taxes care deeply about transactions. Not "risk," not "stress," not "how responsible you were." Taxes mostly care about when you *realize* gains and losses (i.e., lock them in with a sale or expiration), and what *kind* of gains and losses they are. The most important dividing line for many investors is the blunt, expensive one: *short-term* versus *long-term* capital gains.

The two clocks: long-term investing vs. short-term tax math

Long-term capital gains generally apply when you hold an asset for more than a year. Short-term capital gains usually apply when you hold for a year or less. The punchline is simple: short-term gains are typically taxed at higher rates because they're usually treated like ordinary income. Long-term gains often get preferential rates. You don't need to memorize brackets to understand the incentive: if you can choose between "taxed like your paycheck" and "taxed more gently," you'd prefer the gentle option.

Here's the catch: hedges often live on a short clock. Many protective trades-buying puts, rolling options, trimming and re-hedging, selling covered calls to fund protection-reset the tax calendar again and again. The portfolio can still be "long-term" in spirit, but parts of its return stream may get pulled into the short-term world, where taxes bite harder.

Even worse, hedging can change not just the rate but the *timing*. Timing matters because taxes are essentially an interest-free loan you're either receiving (when you defer) or giving (when you accelerate). A hedge that forces you to recognize gains sooner can shrink your compounding engine, even if your pre-tax performance looks fine on paper.

A quick gut-check: why ``tax drag" feels invisible until it doesn't

The most common emotional mistake is to treat taxes as something you "settle up later." But hedging is often about *survival during bad periods*, and taxes have a nasty habit of showing up when you least want another demand on cash.

Consider a simple mindset trap:

- You buy insurance (a put) and feel prudent.
- The market rallies, your put loses value, and you shrug—"That's the premium."
- You do it again next quarter.
- Meanwhile, your other trades (or the way the hedge interacts with your holdings) generate realized short-term gains.

Nothing feels dramatic. Then your tax return reveals that a chunk of what you thought was "portfolio growth" was actually "portfolio growth taxed at the least-friendly rate." The hedge didn't just cost premium; it changed the character of your returns.

This is why two investors can run "the same" hedge and end up with very different results: one manages taxes as part of the design, the other treats taxes as paperwork.

Capital gains character: the simplest version (and why hedges complicate it)

For plain stock or ETF investing in a taxable brokerage account, the story is straightforward:

- Sell after more than a year: likely long-term capital gain or loss.
- Sell within a year: likely short-term capital gain or loss.

Options and other derivatives complicate this in three main ways:

1. *Options are time-bound.* Many expire in weeks or months. That naturally pulls results toward short-term.
2. *Options interact with the underlying.* A hedge on a position can affect when gains/losses are recognized on that position, or whether losses get deferred.
3. *Some contracts live under special tax regimes.* Certain index options and futures can fall under rules that don't behave like regular stock sales.

You don't need a tax degree to hedge intelligently, but you do need to respect one principle: *tax rules were not written to reward "cleverness."* They were written to prevent people from manufacturing losses, deferring gains indefinitely, or pretending they still have risk when they don't. Hedging pushes on exactly those pressure points.

The ``insurance premium" isn't always the premium

When you buy a protective put, the obvious cost is the premium paid. But the full economic cost can include:

- *Tax character shift:* gains that might have been long-term become short-term because of how often you trade or how the hedge is structured.
- *Loss timing restrictions:* a loss you expected to "count this year" gets deferred into the future under anti-abuse rules for offsetting positions.
- *Wash-sale surprises:* a loss gets disallowed because your trades look, to the tax code, like you sold something at a loss and bought it right back.

It's the difference between paying for an umbrella and paying for an umbrella that also quietly increases your health insurance deductible. You still got the umbrella. But the total bill changed.

Short-term gains: the special danger for hedgers

A long-term investor's superpower is patience. The tax code, in its own limited way, rewards that patience. Hedging can accidentally trade it away.

Common hedging behaviors that can tilt outcomes toward short-term taxation include:

- *Frequent rolling.* Buying puts every month, selling before expiry, rebuying.

- *Tactical de-risking*. Selling into dips, buying back quickly, then repeating.
- *Option overwriting*. Selling calls repeatedly for income, especially if you're managing strike/expiry actively.

None of these is "wrong." The point is simply that if you're going to run an active hedge, you should assume the IRS will treat you like an active participant. If your hedge is a treadmill, your taxes may start to look like treadmill taxes.

The wash sale rule: when a loss ``doesn't count" (yet)

Wash sales are the rule that stops investors from selling an investment at a loss and immediately buying it back just to harvest the tax benefit while keeping essentially the same position.

The basic idea: if you sell a security at a loss and then buy the same (or, in many cases, a "substantially identical") security within a window around that sale, the loss may be disallowed and effectively added to the cost basis of the replacement position. Translation: the loss isn't always gone forever, but it doesn't help you *now*, and "now" is often when you wanted it.

Hedgers run into wash-sale headaches because:

- Options can be treated as contracts to buy or sell a security, and those contracts can be "close enough" to trigger wash-sale logic.

- You might close a losing hedge and reopen a very similar hedge quickly—exactly the behavior wash-sale rules are designed to police.

Two practical lessons fall out of this:

1. *Tax-loss harvesting plus hedging requires choreography.* Random, reactive trading is where wash sales thrive.
2. *"Substantially identical" is a facts-and-circumstances monster.* It doesn't always announce itself clearly. Conservative planning beats clever guessing.

Straddles: the rulebook's way of saying ``you can't have it both ways"

A *straddle* (in tax language) broadly refers to offsetting positions-where one position's gains tend to cancel the other's losses. Many hedges are, by design, offsetting. That's the point. But the tax code doesn't love offsetting positions because they can be used to lock in tax losses while keeping economic exposure.

So there are "straddle rules" that can do things like defer losses on one leg if you still have an offsetting position open. In plain English: you don't always get to claim the losing side right away if you're still holding the winning side that would prove you never really left the risk behind.

This matters for hedgers because a common instinct is:

> "I'll take the loss on the hedge for taxes, and I'll keep the appreciated stock because I'm a long-term investor."

Tax rules often respond with:

> "Nice try. If those positions are tied together economically, the timing of that loss may not be up to you."

If you've ever wondered why sophisticated investors talk about hedge design with the wary tone of someone describing a kitchen remodel ("It started simple, then the walls came off"), straddle rules are one reason.

Covered calls: income, regret, and the calendar you didn't know you were resetting

Selling covered calls can feel like getting paid for patience. You own the stock anyway; you sell someone else the right to buy it from you at a certain price; you collect a premium. In calm markets it can look like "extra yield."

Taxes add two layers of realism:

- Option premium income can be short-term in nature depending on how the option position closes (expires, bought back, assigned).
- Covered-call writing can interact with other tax rules—especially when it's combined with protective puts or when it affects whether your stock position is considered meaningfully "hedged."

There are also special concepts like "qualified covered calls" that can matter for tax treatment in certain cases. The details get technical fast, but the intuition is simple: if the call you sell is so aggressive (or structured in a certain way) that it behaves less like "income on a long position" and more like "I've boxed myself into a near-riskless trade," the tax system may start treating the whole setup differently.

A hedger's takeaway: covered calls are not just an options strategy; they are a *tax strategy whether you want them to be or not*. If you sell calls mechanically, you may be mechanically manufacturing a certain type of taxable result.

Section 1256 and the strange, friendlier world of 60/40

Not all derivatives are taxed the same way. Some contracts fall under special rules (commonly associated with certain futures and certain index options). One famous feature of this regime is a "60/40" blend: a portion treated as long-term and a portion treated as short-term, regardless of how long you held the contract.

That can be a meaningful advantage for active hedgers, because it partially escapes the "everything is short-term because I roll often" trap. Another feature you may encounter is "mark-to-market," where positions are treated as if they were sold at year-end for tax purposes, even if you didn't actually close them.

The correct emotional posture here is neither excitement nor fear—it's respect. A favorable rule is still a rule. If your hedge uses instruments that can qualify for this treatment, it may reduce the tax bite of frequent adjustments, but it can also change when gains and losses show up on your return.

If you're choosing between two ways to hedge the same risk—one that tends to generate pure short-term outcomes and one that may receive blended treatment—taxes can be the tiebreaker that turns "roughly equivalent" into "materially different."

The ``tax friction" test: a hedge is only as cheap as its after-tax cost

Investors often compare hedges by premium: "Strategy A costs 2% per year, Strategy B costs 1.5%." That's a start, but it's not the finish.

A more realistic comparison is: *How much protection do I get per dollar of after-tax cost?*

Why after-tax? Because two hedges with the same pre-tax drag can have very different after-tax drags:

- One hedge produces occasional long-term gains (or blended treatment), keeping more of the upside you do earn.
- Another hedge produces frequent short-term gains that siphon off a larger share of any gains you realize, even in years when you didn't feel like you "made money."

Taxes can also cause a strategy to behave differently across account types and income situations. A hedge that is tolerable for someone in a low tax bracket can be brutally inefficient for someone in a higher bracket, even if the market results are identical.

The hedger's temptation: ``I'll take the loss, keep the gain, and call it smart"

There's a particular kind of trade that feels emotionally perfect: you take a tax loss on the hedge (because it's down), you keep holding the appreciated asset (because you believe in it), and you walk away thinking you engineered a free lunch.

Tax rules are built to detect lunches that look too free—especially when you haven't really changed your economic exposure. If your setup is designed so that you can claim losses now while keeping gains deferred indefinitely, you're walking into the part of the code that was written by people whose job is to prevent exactly that.

The practical way to stay out of trouble isn't to avoid hedging; it's to avoid *pretending hedging doesn't change your tax profile.* Once you accept that hedging and taxes are intertwined, you can design with that reality rather than collide with it.

A sane way to think about it: pick what you're optimizing

Taxes aren't the only objective. If your hedge prevents you from panic-selling during a crash, it can be worth paying some tax friction. The danger is paying the friction *by accident*.

It helps to choose which "win" you're aiming for:

- *Max after-tax return:* minimize short-term realizations; prefer structures with favorable character where appropriate.
- *Max behavioral protection:* accept some tax cost to keep the plan survivable in drawdowns.

- *Max simplicity:* fewer moving parts, fewer nasty surprises, even if it's not the most tax-efficient on paper.

You can't perfectly optimize all three at once. But you can avoid the worst outcome: complexity *and* high tax drag *and* fragile behavior.

Actionable takeaway

Before placing a hedge in a taxable account, answer these three questions in writing (one sentence each), and keep the note with your trade plan:

1. *What is the holding-period goal?* (Am I trying to preserve long-term capital-gain treatment on my core holdings, or am I accepting more short-term activity?)
2. *What is the realization plan?* (Exactly when do I close, roll, or let expire—and what will trigger a change?)
3. *What rule might bite me?* (Wash sale, straddle loss deferral, covered-call complications, or special contract treatment—pick the one most likely and confirm it with a tax professional or high-quality tax guidance before scaling up.)

If you can't answer those cleanly, the hedge may still reduce market risk—but it's likely increasing *tax* risk in a way that won't feel real until it's expensive.

7.2 Rules of the Road: Hedging in IRAs and 401(k)s

The first time someone tries to "do the sensible thing" inside a retirement account—trim risk, add protection, maybe short a little something—they usually run into the same experience: the trade ticket feels like a locked door. No margin. No short selling. Options permissions that read like a restaurant menu where half the items are grayed out. And it's not because your broker is being petty. Retirement accounts were built to be wealth *vaults*, not trading *workshops*. The rules are designed to prevent you from pledging, borrowing, or levering up the very money the government is giving special tax treatment.

That design choice has a personality. A taxable brokerage account is a wide-open highway with speed traps; a retirement account is a guarded park with signs about what you're not allowed to feed the wildlife. You can still get plenty done—but you do it by working with the constraints instead of headbutting them.

Two kinds of ``can't": what the tax code forbids vs. what your custodian won't allow

When people say, "You can't do that in an IRA," they're often blending two separate limits into one sentence:

- *Legal limits* (the tax rules): prohibited transactions, borrowing, pledging the account, and similar "self-dealing" behavior that can blow up the account's tax-advantaged status.

- *Policy limits* (the broker/custodian rulebook): which option strategies are permitted, what level of approval you can get, and what collateral they require.

This distinction matters because the consequences are wildly different. If your broker won't let you place a trade, you get an error message and mild annoyance. If you manage to cross a prohibited-transaction line, the punishment can be catastrophic: the IRA can be treated as no longer an IRA as of the first day of that year, which can trigger taxes (and potentially penalties) as if the account were distributed at fair market value. That's not "oops." That's a financial house fire.

So the game isn't "find clever loopholes." The game is "build protection that stays boring in the eyes of the IRS and your custodian."

Why margin and short selling are usually out

In ordinary brokerage accounts, margin is just a tool: borrow against your securities, amplify returns (and losses), and hope the market cooperates. In an IRA, borrowing and leverage start to look like you're using retirement assets for your own benefit in ways the rules were written to prevent. Many prohibited-transaction concepts orbit around lending/extension of credit and using plan assets as security. Even if you personally feel conservative, the structure itself makes regulators nervous for good reason: leverage turns a "long-term savings plan" into something that can be forced into liquidation at the worst possible moment.

Short selling runs into the same wall. Shorting typically involves borrowing shares and posting collateral; it's margin by another name. Add in the practical reality—retirement accounts can't have negative cash balances at most custodians—and short selling becomes a non-starter.

The simplest way to remember it: retirement accounts are meant to grow in a tax-sheltered environment, not to become a personal line of credit with a brokerage logo.

Options in retirement accounts: not impossible, just fenced in

Options are where many investors find workable, elegant hedges—because a put is literally insurance in contract form. The surprise is that retirement accounts often allow *some* options, but rarely allow the kind of options behavior that can create open-ended obligations.

In practice, many custodians will approve "defined-risk" or "fully-collateralized" strategies in IRAs, such as:

- *Buying calls and puts* (you pay the premium; your risk is capped at that premium).
- *Covered calls* (you already own the shares, so if assigned you deliver stock you have).
- *Cash-secured puts* (you keep enough cash to buy the shares if assigned).
- *Certain spreads* (where the maximum loss is limited and collateralized).

What is commonly restricted are trades with theoretically unlimited risk or margin dependence:

- Naked short calls or naked short puts.
- Many margin-based multi-leg positions.
- Anything that could create a deficit the account can't legally carry.

The emotional shift is important. In a retirement account, you don't "sell volatility" like a professional market maker; you buy protection or you use options to reshape outcomes with clearly bounded risk.

The IRA ``third rail": prohibited transactions and why they're so severe

If there's one topic that deserves a little healthy fear, it's prohibited transactions. People assume retirement-account mistakes are like parking tickets—pay a fee, move on. Some are. This one can be more like getting your driver's license revoked.

The prohibited-transaction framework is aimed at preventing "self-dealing"—using retirement money in ways that benefit you personally right now, outside the intended retirement purpose. Borrowing from the IRA, pledging IRA assets as security, or otherwise using the account to support personal obligations can trigger the rules. The nightmare scenario is that the account loses its protected status and is treated as distributed, which can create a big tax bill and, depending on age and circumstances, additional penalties.

This is why certain "clever" hedging ideas are dangerous even if they sound responsible:

- Using the IRA as collateral for a loan ("It's my money, I'm just backing a credit line").
- Trying to simulate margin-like behavior through side agreements.
- Anything that looks like the account is guaranteeing or supporting your personal finances.

A good rule of thumb: if a strategy requires a sentence that starts with "Technically...", it probably doesn't belong in an IRA.

401(k)s: the plan menu problem

If IRAs feel fenced in, 401(k)s can feel like shopping at a convenience store. The limiting factor is usually not the tax code; it's the plan's investment menu. Many plans offer mutual funds, target-date funds, maybe a brokerage window if you're lucky, and sometimes company stock. Options trading is typically not available inside the plan itself.

That doesn't mean you can't hedge your *household* portfolio. It means you hedge at the portfolio level using the accounts you can actually maneuver. If your 401(k) is 60% of your net worth and it's 100% stock funds, the practical solution may be to adjust risk in the 401(k) with the choices you have (more bonds, different allocation) and reserve tactical hedging tools for an IRA or taxable account—without turning the whole retirement plan into a high-maintenance machine.

One underappreciated truth: the best retirement-account hedge is often not a derivative. It's a thoughtful asset allocation that prevents you from needing a derivative.

The ``steering wheel" approach: hedge the whole portfolio, not each account

Many investors make the same mistake in reverse: they treat each account like it must be self-contained and perfectly hedged. That leads to bad engineering—like buying protection in an IRA because "that's where the stocks are," even if the IRA's trading rules make the hedge clumsy or expensive.

A cleaner approach is to think of your accounts as parts of one vehicle:

- The 401(k) might be the engine: big, steady, not designed for stunts.
- The IRA might be the steering wheel: more flexibility, still governed by safety rules.
- The taxable account might be the suspension: where you can fine-tune, but you feel every bump (including taxes).

The hedge doesn't need to sit in the same account as the risk, as long as the total exposures line up. If your household portfolio is effectively "long U.S. stocks," you can offset some of that with an IRA-held put on a broad index ETF—if your custodian allows it—without pretending the IRA and 401(k) are separate universes.

Inside the fences: practical hedging tools that often work

1) Long puts on broad market exposure (if your IRA allows options).

A long put is the purest "retirement-friendly" hedge because it's prepaid insurance. You pay the premium; you cannot lose more than that premium. There is no borrowing, no margin call, no hidden obligation that can exceed the cash you put in.

The catch is psychological and mathematical: long puts are a recurring cost. In a taxable account, you might care about the tax character of frequent option trades. In an IRA, the tax character is less of a day-to-day concern because the account is tax-advantaged. That's the upside. The downside is you can get comfortable "buying insurance forever" without noticing that the drag is real money leaving the compounding engine.

This is where retirees and near-retirees sometimes make a smart trade: accept a known, bounded insurance cost during the years when a crash would do the most damage.

2) Covered calls as a speed limiter.

If you already own a broad stock ETF in an IRA and your custodian allows covered calls, selling calls can reduce volatility and generate premium. It's not crash insurance. It's more like agreeing to cap some upside in exchange for getting paid today.

The danger is forgetting what you sold. In a roaring bull market, covered calls can make you feel like you're always "a little behind," because you keep handing away the explosive upside. That's not a flaw; it's the contract. In retirement accounts, covered calls can be a reasonable way to turn down the portfolio's emotional temperature—if you accept that you're swapping some upside for steadier outcomes.

3) Cash-secured puts as a disciplined entry tool.

Cash-secured puts can look like a hedge because they generate income, but they're really an "I'd be willing to buy at a lower price" commitment. In an IRA, this can be useful if your plan is to hold a core index exposure and you want a more structured way to add on dips—while being honest about the obligation by holding the cash.

The gotcha is that selling puts increases downside exposure if the market falls hard. If the goal is protection, don't confuse "premium collected" with "risk reduced." Premium is not a shield; it's a payment for taking on a specific risk.

4) Defined-risk spreads (when permitted).

Some custodians allow limited spreads in IRAs. These can be a way to reduce the cost of hedging by capping both the protection and the premium outlay. Think of it as buying insurance with a deductible and a policy limit. You pay less, but the protection is partial.

Spreads can be excellent tools for making hedging affordable. They can also become a hobby. If you find yourself adjusting spreads weekly, you're not hedging anymore—you're running a small options desk from your retirement account.

5) Inverse ETFs: the ``no-options" hedge with a personality.

If your retirement account doesn't allow options (common in 401(k)s and some IRAs), inverse ETFs are often mentioned as an alternative: they aim to move opposite a benchmark. They're easy to buy and sell like a normal fund.

But inverse and leveraged ETFs come with their own quirks, especially over longer holding periods, because daily rebalancing can cause returns to drift away from the simple "opposite of the index" story. Used briefly and deliberately, they can be a functional tool. Used as permanent hedges, they can become expensive in ways that don't feel like a premium but behave like one.

A good mental model is to treat them like a tool you take out during a storm—not a foundation you build the house on.

Asset location: putting the ``high-tax behaviors" where they hurt less

One of the most useful practical concepts for hedgers is *asset location*: not what you own, but *where* you own it.

Even if you never become an options expert, you can still win by placing "tax-inefficient" activity in tax-advantaged accounts when possible. Hedging is often tax-inefficient in taxable accounts because it tends to generate short-term results and frequent realizations. A retirement account can shelter some of that churn.

This is not a license to turn the IRA into a casino. It's simply good plumbing:

- Put the most frequently adjusted hedges in the accounts where those adjustments don't create immediate tax paperwork.
- Keep long-term holdings that you want to enjoy at favorable long-term capital gains rates in taxable accounts when it fits your broader plan.

Of course, the constraints bite again: if your 401(k) won't let you do anything except hold mutual funds, then your "location" decision might be more about where you hold bonds, where you hold equities, and where you allow yourself to do tactical trades.

The cash problem: hedging can create liquidity needs in the wrong place

Retirement accounts are designed to discourage casual withdrawals. If you're under age 59½, pulling money out can trigger additional taxes and penalties unless an exception applies. Even above that age, a withdrawal is still a taxable event for many account types.

So any hedge that can create a sudden need for cash—assignment on a short put, for example—needs to be structured so the IRA can handle it internally without forcing distributions. That's one reason custodians prefer "cash-secured" and "defined risk" strategies: they don't create surprise liabilities.

The practical takeaway is boring but powerful: in retirement accounts, every strategy should have a clear collateral story. If you can't point to the cash or shares that cover the worst case, you're one unpleasant surprise away from an unnecessary tax event.

Employer stock inside a plan: sometimes the hedge is the exit ramp

Some 401(k)s include employer stock. That can turn into concentration risk—your paycheck and your retirement money tied to the same company. Hedging employer stock inside the plan is often not available. The real-world solution may involve distribution planning (when eligible) rather than derivatives inside the plan.

This is a place where tax rules can become opportunity *or* landmine depending on circumstances, and it's exactly where professional advice earns its keep. The general lesson for a long-term investor is simpler: when the risk is a concentrated position inside a restrictive wrapper, the hedge may not be a trade—it may be a carefully planned reduction of exposure using the plan's allowed paths.

A reality-based permission checklist (the one investors wish they'd used sooner)

Before you design a hedge in an IRA, answer these questions like a pilot running a pre-flight check:

1. *What does my custodian explicitly allow in this account?* (Options level, spreads, assignment handling, settlement rules.)
2. *Is every obligation fully collateralized with cash or owned shares?* (If not, don't do it in an IRA.)

3. *Could this be construed as borrowing, pledging, or self-dealing?* (If yes or even "maybe," stop and get professional tax guidance.)
4. *If the market gaps down overnight, what happens tomorrow morning?* (You want "nothing dramatic.")

The best hedges in retirement accounts share a personality: they're robust, boring, and hard to break.

Actionable takeaway

Call (or message) your IRA custodian and ask for their *written* options permissions list for IRAs—what levels they offer and which specific strategies are permitted (covered calls, long puts, cash-secured puts, spreads). Then write a one-paragraph "retirement-account hedging policy" for yourself: which strategies you will use, what maximum premium budget you'll spend per year, and what you will *never* do (anything margin-like, anything not fully collateralized, anything that could force a distribution). Keep that policy next to your investment plan, and treat it like a safety railing—not a suggestion.

7.3 The Retirement Red Zone: Conquering Sequence Risk

A market crash at 30 is a bruise. A market crash at 65 can be a broken hip.

That's not drama; it's arithmetic. When you're still working, your paycheck is a built-in recovery machine. You keep buying shares—often at newly discounted prices—and time does the heavy lifting. When you're retired, withdrawals run the movie in reverse: you're forced to *sell* shares to fund life, even when prices are ugly. The same bear market that looks like a temporary setback on an accumulation chart can look like a trapdoor on a retirement chart.

Sequence of Returns Risk is the risk that the *order* of returns matters more than the *average* return once withdrawals begin. Two portfolios can earn the same average annual return over 30 years; the one that suffers its bad years early can run out of money, while the one that suffers its bad years late can survive comfortably.

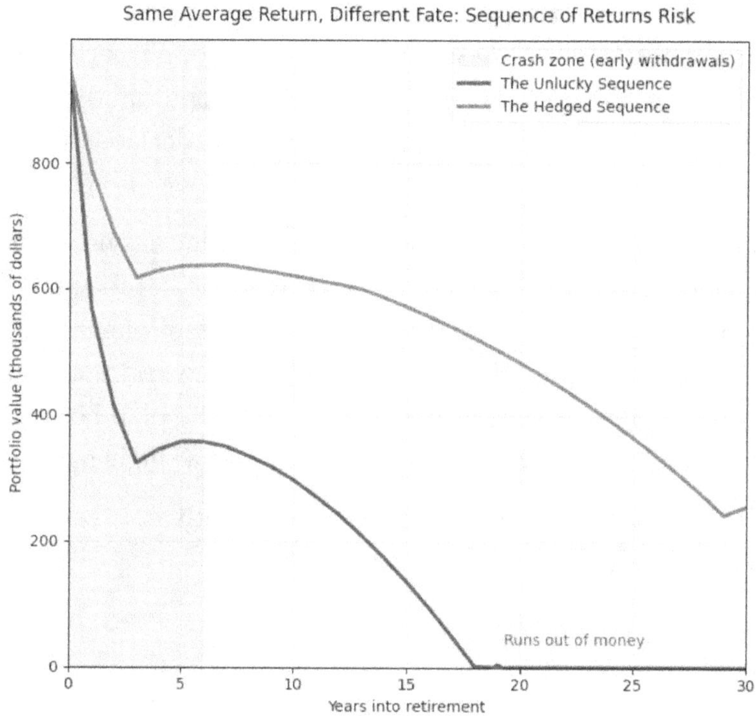

Illustration: fixed 50*kwithdrawalsfrom*1.0M; both paths have identical average annual return (4.9%).

If that chart makes you a little uncomfortable, good. Retirement planning is one of the few places in personal finance where discomfort is a feature, not a bug. The "unlucky" retiree didn't do anything reckless. They didn't buy meme stocks. They didn't forget to diversify. They simply retired into a bad decade and had the audacity to keep paying the electric bill.

Why early losses are so toxic when withdrawals start

The problem is not volatility by itself. The problem is volatility *plus* withdrawals. When the portfolio is down, every dollar you withdraw forces you to sell *more* shares to raise the same cash. Those shares are gone forever. They don't participate in the eventual recovery. A retirement portfolio can die from "share count bleed" long before the headlines declare the bear market over.

A quick example that mirrors real life: imagine a retiree with \$1,000,000 and a \$50,000 annual spending need. If the market drops 30% early, the \$50,000 withdrawal is no longer 5% of a million; it's more like 7% of \$700,000. The withdrawal rate silently spikes at exactly the wrong time. The plan becomes self-sabotaging without the retiree changing a single habit.

This is why retirement isn't just "investing with more caution." It's investing with a new villain: *bad timing*.

The ``red zone'': the 5--10 year window where risk is least forgiving

The years right around retirement—the last few years of earning, and the first several years of withdrawals—are the red zone. Your human capital (future earnings) is shrinking or gone, and your portfolio has just become your paycheck. A major drawdown in this window can permanently lower the paycheck your assets can sustainably produce.

The counterintuitive part is that the red zone isn't always the day you stop working. It can start earlier. A 58-year-old planning to retire at 63 is already in the zone if the plan relies on that portfolio value being intact. "I'll just work longer" might be true, but it's not always easy, available, or emotionally tolerable—especially if the downturn coincides with layoffs, health issues, or caregiving responsibilities. Sequence risk is personal because life is personal.

The goal is not to avoid drawdowns; it's to avoid forced selling

Most retirees don't need a portfolio that never drops. They need a portfolio that doesn't *force* them into irreversible decisions at bad prices. The hedge, in the retirement context, is less about winning and more about buying *time*: time for markets to normalize, time to adjust spending gradually instead of violently, time to keep the plan intact.

That shift in goal changes what "good hedging" looks like. The best retirement hedge often has these traits:

- It creates cash (or cash-like stability) exactly when stocks are ugly.
- It is simple enough to run when you're stressed.
- It avoids unnecessary tax and account friction that could force distributions or extra selling.

In other words, it behaves like a shock absorber, not a day-trading system.

The bond tent: building a temporary ceiling over your head

A *bond tent* is one of the most humane ideas in retirement planning because it admits something quietly true: you don't need the same risk profile at every age. The concept is to raise your bond allocation as you approach retirement (the tent goes up), then gradually lower it later in retirement (the tent slopes down) once the most fragile early years have passed.

Why can this work? Because bonds (or cash-like assets) can fund withdrawals in a bear market, allowing you to avoid selling stocks at depressed prices. You're not predicting the market. You're creating a buffer that reduces the chance of selling the family's future at a garage-sale price.

Bond tents get criticized when bonds have a bad year (which they can). But the retiree's use of bonds is not "bonds always go up." The purpose is *liability matching*: having a pile of money that is meant to be spent in the next several years, so the stock portion can be left alone to recover.

A practical way to think about it: you're not "investing conservatively." You're pre-paying for the right to be patient.

The cash wedge: the simplest hedge that actually gets used

The least glamorous retirement hedge is also one of the most effective: a cash wedge (or short-term reserve). Not 0% cash, not 30% cash forever—just enough to cover a defined period of spending without touching stocks.

Cash has an obvious drawback: it doesn't compound like equities. But in the red zone, compounding is not the only job. Cash has a psychological superpower: it prevents panic. A retiree with two years of spending in cash can watch a bad market and still buy groceries without feeling like each purchase is a tiny act of sabotage.

Cash also plays nicely with retirement-account constraints. You don't need margin. You don't need options approvals. You don't need complex trades that might be restricted by a 401(k) menu or an IRA custodian. You just need a plan for replenishment: when markets are healthy, refill the wedge; when markets are sick, spend the wedge.

This is hedging you can execute while half-asleep, which is a real criterion when markets are screaming.

Protective puts near retirement: expensive, powerful, and surprisingly honest

Buying a protective put—an option that increases in value when the market falls—can be a direct antidote to sequence risk. It can also feel like lighting money on fire in calm markets, because the put loses value if the market rises or drifts.

The honest way to use protective puts in the red zone is to treat them like insurance on a house you cannot rebuild. If your portfolio is large enough and your spending needs are rigid, the question becomes less "Is the premium annoying?" and more "What is the cost of being wrong without it?"

The retiree use-case isn't to hedge forever. It's to hedge *when the plan is fragile*. A temporary program—say, maintaining downside protection for a few years around the retirement date—can be a bridge. The bridge does not need to last your whole life; it needs to get you across the ravine.

Retirement accounts can be a natural home for this kind of activity when permitted, because the churn of rolling options doesn't create the same annual tax paperwork. But the operational constraint matters: you need an IRA custodian that allows long puts (many do), and you need a rules-based approach so you don't improvise under stress.

Spending rules: the hedge you control with a pen, not a trading platform

A subtle tragedy in retirement is watching someone with plenty of assets get cornered by a too-rigid spending rule. If your withdrawal plan is "always take $X no matter what," you are effectively promising to sell shares regardless of price. That's not discipline; it's a blindfold.

A better approach is to build flexibility into spending—a pre-commitment that you will tighten the belt in down markets and loosen it in up markets. This doesn't mean living on ramen in a recession. It means distinguishing between spending that is truly fixed (housing, healthcare, basic living) and spending that can breathe (travel, gifts, discretionary upgrades).

When retirees adopt even modest dynamic spending rules, they often reduce sequence risk dramatically without buying a single option. It's not exciting, but it's powerful: the market cannot force you to sell what you refuse to spend.

Asset location and taxes: the retirement twist that sneaks up on people

In retirement, what you sell matters as much as what you own. A household might have:

- taxable accounts (where selling can generate capital gains taxes),
- traditional tax-deferred accounts (where withdrawals are typically taxed as ordinary income),

- Roth accounts (where qualified withdrawals can be tax-free).

Sequence risk gets nastier when a bear market coincides with a tax-inefficient withdrawal pattern. Selling the "wrong" asset in the "wrong" account can increase the number of shares you must liquidate to fund the same spending. Taxes, in that moment, behave like an extra withdrawal.

This is why retirees often benefit from a coordinated drawdown plan—deciding in advance which accounts fund which years, and how market declines change that ordering. It's also why hedges that create complexity in taxable accounts can be less attractive near retirement: the last thing you want is a protection program that accidentally increases realized income in a year when you're trying to manage tax brackets, Medicare surcharges, or other income-linked thresholds.

The best retirement hedges are the ones that still make sense after you add taxes to the picture.

The ``bucketing" mindset: time-segment your money

Many retirees find it easier to stick to a plan when money has jobs. A classic approach is bucketing:

- *Bucket 1:* near-term spending (cash and short-term, high-quality bonds)
- *Bucket 2:* mid-term stability (intermediate bonds, balanced exposure)
- *Bucket 3:* long-term growth (equities)

This is not magical accounting. It's behavioral engineering. When stocks crash, Bucket 1 is the oxygen tank. It prevents "sell low to live" from becoming your default response.

A bucket plan also makes rebalancing feel less like gambling. You're not guessing bottoms; you're replenishing the spending bucket when markets have been kind, and refusing to drain the growth bucket when markets are cruel.

The retiree's stress test: a simple way to see if you're in danger

Sequence risk hides because average return looks comforting. So test the plan in the only way that matters: start with a bad decade.

Ask: what happens if the first two years of retirement include a 30% equity drawdown and a sluggish recovery? Do you have enough stable assets or hedges to avoid selling stocks aggressively? If the answer is "I guess I'd just sell anyway," the plan is not a plan—it's a hope.

A good stress test doesn't require fancy software. Even a rough spreadsheet with conservative assumptions can tell you if your first five years are survivable without heroics. If your plan only works when markets cooperate, you don't have a retirement plan; you have a market forecast.

Putting it together: what ``conquering" sequence risk actually means

Conquering sequence risk is less about outsmarting the market and more about refusing to be cornered. You're building multiple escape routes:

- a buffer (cash wedge or bond tent),
- a rule (dynamic spending),
- possibly an explicit hedge (protective puts or defined-risk structures),
- and a withdrawal plan that respects taxes and account constraints.

Notice what's missing: predictions. The strongest retirement defenses don't require you to be right about the next year. They only require you to be prepared for the next year to be unpleasant.

Actionable takeaway

Write a one-page "Red Zone Policy" for the first 5–10 years around your retirement date:

- Define a minimum spending floor (must-pay expenses) and a flexible spending layer (nice-to-have).
- Choose a buffer size (e.g., 18–36 months of spending) and where it will live (cash, short-term bonds, or a bond tent allocation).
- Set a rule for bear markets: "If equities are down more than __% from their peak, fund spending from the buffer and pause stock sales."
- If you plan to use options, write the simplest possible rule for buying and rolling protective puts (what index/ETF, how far out, what budget), and confirm your IRA custodian permissions before you need them.

Then do the unglamorous part: rehearse it once. Open last decade's chart, imagine retiring into the ugliest year, and walk through exactly which assets you would sell first. The goal is to make the first real bear market of retirement feel like executing a plan—not improvising a rescue.

7.4 All Eggs in One Basket: Managing Concentrated Stock

There's a special kind of anxiety that comes from opening your brokerage app and realizing your financial future has a logo.

For a lot of people, concentrated stock isn't a reckless bet-they didn't "go all in" on a whim. It's what happens when life is generous in a lopsided way: employee stock that piled up quietly over a decade, a founder's stake that never got diversified, an inheritance that arrived as a single ticker, or options and RSUs that turned into real shares faster than anyone emotionally processed. On paper you're wealthy. In practice, you're one earnings call away from a very bad year.

And the cruel part is that the obvious fix-"just sell and diversify"-often feels impossible. Sometimes it *is* impossible due to blackout windows, insider rules, or restricted stock. Sometimes it's possible but psychologically brutal because selling triggers a tax bill big enough to feel like lighting a portion of your net worth on fire. Concentration risk is the only risk that can make a person feel trapped by success.

Why concentrated stock is different from ``a risky portfolio"

A diversified portfolio can have drawdowns, but it usually doesn't have a single point of failure. Concentrated stock does. Company-specific risks-fraud, litigation, regulation, a product miss, a scandal, a CEO departure, an industry shift-can permanently impair one company even while the overall market recovers. Index investors get to be wrong about a few companies and still retire. A concentrated holder doesn't have that luxury.

There's also a timing problem. With a diversified portfolio, you can often reduce risk gradually with rebalancing. With concentrated stock, the biggest loss can come *before* you get the chance. The market doesn't wait for your tax planning appointment.

So the real objective is not "maximize returns." It's: *reduce the chance of a single-company disaster while managing taxes, rules, and human emotions.* That's a three-body problem. You need a plan that works in the real world, not just in a spreadsheet.

The three fences that trap concentrated holders

Most concentrated stock plans fail because they ignore one of three fences:

Fence 1: Taxes. A highly appreciated position is a loaded spring. Selling releases the spring and the tax bill arrives with the force you've been deferring for years. The bigger the embedded gain, the more tempting it is to postpone-until the market makes the decision for you.

Fence 2: Legal and compliance constraints. Employees and executives can face blackout periods, insider trading rules, and restrictions on when and how they can trade. Affiliates and holders of restricted/control securities may also face volume limits and "manner of sale" requirements under SEC Rule 144, which can slow diversification to a drip even when you're desperate to move faster.

Fence 3: The tax code's suspicion of "perfect hedges." When you try to lock in gains without selling-by using short sales, forwards, or other offsetting trades-the tax code can treat it like you sold anyway. That's the constructive sale problem: if you eliminate substantially all risk on an appreciated position, the IRS may decide you've effectively cashed out and should pay tax now.

The result is frustrating: selling can be expensive, but "not selling" can be dangerous, and "hedging without selling" can still trigger tax consequences if you overdo it. This is why concentrated stock requires tailored solutions rather than generic hedging recipes.

Start with triage: how concentrated are you, really?

Before tactics, you need a diagnosis. Concentration isn't a moral failing; it's a measurement.

A practical way to define "too concentrated" is not a single magic percentage. It depends on your job stability, your spending needs, and whether the stock is also correlated with your income (employees often have both tied to the same company). But a simple triage framework helps:

- *Mild concentration:* uncomfortable, but a 50% stock drop wouldn't change your life plan.

- *Serious concentration:* a 50% drop would force major changes (retirement date, housing, family plans).
- *Critical concentration:* a 50% drop could be financially devastating.

If you're in the serious or critical category, the "cost" of protection should be compared to the cost of a ruined plan, not to the cost of perfection. This is where people often surprise themselves: they'll spend $2,500 a year on home insurance without blinking, but they'll agonize over a hedge that costs a similar fraction of their portfolio because the hedge feels like admitting vulnerability.

The cleanest solution is still the most boring one: staged selling

If you *can* sell, staged selling is usually the backbone of the plan. It's not glamorous, but it's brutally effective. You decide in advance how much you'll sell per month or per quarter, and you stick to it through good markets and bad. The point is to reduce the decision-making surface area. Concentrated holders don't need more cleverness; they need fewer moments of panic.

Staged selling also helps with a common psychological trap: waiting for the "right price." The right price is often a mirage. A rules-based program turns "market timing" into "risk management," which is what you actually want.

For employees and insiders, this staged approach often pairs naturally with a trading plan structure designed to reduce insider-trading risk and decision stress (subject to legal and compliance requirements). The big benefit isn't just legal safety; it's that the plan keeps working when emotions would otherwise sabotage it.

When selling is hard: the collar as a sleep-at-night strategy

Now for the tool that exists specifically for the concentrated-stock problem: the *collar*.

A collar is two options wrapped around your stock position:

- You *buy a put* below today's price. That put is your floor: if the stock collapses, the put gains value and helps limit losses.
- You *sell a call* above today's price. That call is your ceiling: if the stock rockets upward, you may have to sell at the call strike (or you may have to manage assignment), giving up some upside.

You're literally trading away a chunk of future upside to pay for downside protection today. It's not free money. It's a decision about what kind of regret you prefer:

- Regret A: "I capped my upside and the stock went to the moon."
- Regret B: "I kept all the upside potential and the stock cut in half."

In the concentrated-stock context, Regret A is often survivable. Regret B can be life-altering.

The zero-cost collar (and why ``free'' is a dangerous word)

A *zero-cost collar* is a collar structured so that the premium you receive from selling the call approximately pays for the put you buy. Done properly, you can sometimes create protection with little to no out-of-pocket premium.

This is where investors get excited—and where they should slow down.

"Zero-cost" describes the option premiums, not the true cost. The true cost is the upside you give away. If the stock rallies hard, you may find yourself effectively forced to sell (or manage around assignment), and that can accelerate a taxable event you were trying to postpone. In other words: you didn't pay cash, but you paid with *future flexibility*.

Still, for someone stuck holding a massive position they can't sell right now, a collar can be the first moment of real relief. It converts a terrifying, open-ended downside into a defined range. Humans do better when the worst case is finite.

A collar is not just an options trade; it's a tax and compliance object

Concentrated stock is where the earlier tax themes become very real: the government pays close attention when you appear to "lock in" gains without selling. Under constructive sale rules, certain transactions that substantially eliminate risk on an appreciated position can cause you to recognize gain as if you sold the position. Short sales against the box and certain forward/futures-like arrangements are classic examples of what the rules target.

A vanilla collar does not automatically trigger constructive sale treatment in every case, but concentrated-stock hedging lives close to that line, and details matter: strike selection, expirations, whether the collar is "tight," whether it eliminates too much risk, whether it resembles other targeted transactions. This is why concentrated stock is the moment to stop improvising and coordinate with a tax professional who has seen these structures before. The goal isn't to play chicken with the tax code. The goal is to reduce risk without detonating the tax bill prematurely.

Compliance also matters. If you're an insider, your company may have rules about derivatives on company stock, not just outright sales. Some companies restrict or prohibit certain hedging transactions for executives because hedging can undermine the alignment between management and shareholders. The best collar in theory is useless if it violates your employer's policy.

How to think about collar sizing without drowning in math

The seductive mistake is to try to collar *everything* at once. The practical mistake is to collar too little and call it "risk management."

A more realistic approach is to pick a "protected core" amount-enough shares that if the stock gets hit, your life plan still works. Then decide what portion remains "unprotected upside" that you can tolerate being volatile.

This reframes the decision. You're not trying to build a perfectly engineered derivative structure. You're trying to answer a human question: *How many shares do I need to stop the position from ruining my life if it breaks?*

Once you know that number, the collar becomes a tool, not a philosophy.

Alternative levers: reduce concentration without selling in the obvious way

Collars aren't the only solution, and sometimes they're not the right one. Here are other levers that can matter, depending on your constraints:

Charitable giving of appreciated shares. Donating appreciated stock can reduce concentration while potentially avoiding capital gains tax on the donated shares and producing a charitable deduction (subject to rules and limits). This can be an elegant "two birds" move for charitably inclined investors, especially when paired with a disciplined giving plan rather than a one-time emotional donation.

Using new cash flow to diversify aggressively. This sounds trivial, but it's often overlooked. If you receive bonuses, RSU vesting, or other cash flows, directing *all new money* away from the concentrated stock (and into diversified holdings) is a quiet way to reduce concentration over time without triggering immediate tax on the existing position.

Risk reduction via the rest of the portfolio. If you cannot touch the concentrated position, you can sometimes reduce total household risk by being more conservative elsewhere—more bonds, more cash, less equity exposure outside the concentrated stock. This doesn't solve company-specific risk, but it can keep a single-stock blow-up from becoming a total-net-worth catastrophe.

Plan-level special cases for employer stock. If the concentrated position sits inside an employer plan, distribution mechanics can sometimes create unusual tax planning opportunities, but the rules are specific and the stakes are high. This is where "special case" really means special: it can change the economics dramatically, but it requires careful eligibility and execution.

Exchange-fund style diversification concepts. There are structures in the market that aim to convert concentrated stock into a diversified basket without immediate recognition of gain, often through partnership-style contributions. These tend to be complex, limited to certain investors, and highly dependent on structure and rules. They can be powerful, but they are not do-it-yourself projects.

Each of these approaches can reduce concentration risk, but they come with different mixes of complexity, cost, and control. The right tool is the one that you can actually implement under your real constraints.

The hidden risk: your paycheck is often the same stock in disguise

For employees and executives, concentration isn't just the shares. It's the correlation. If your company hits trouble, you might face:

- a falling stock price,
- reduced bonuses or equity comp,

- layoffs or reduced job security,
- and a shrinking industry opportunity set.

That's why "I can tolerate volatility" is not the right question. The right question is: *Can I tolerate volatility at the same time my income is threatened?* This is where hedging becomes less about optimizing returns and more about stabilizing a family's operating system.

A practical blueprint that respects reality

A concentrated-stock plan that works usually includes these elements-not because they sound nice, but because they solve different parts of the trap:

- *A selling program* for the shares you *can* sell (even if it's slow).
- *A protection program* (like a collar) for the shares you *can't* sell right now but can't emotionally leave naked.
- *A policy check* with employer compliance and trading windows.
- *A tax map* that makes the capital gains consequences explicit, rather than a vague fear.
- *A household allocation adjustment* so your total risk is sane even before the concentration is fully solved.

The best part of doing this well is not higher returns. It's that you get your attention back. Concentrated stock steals mental bandwidth. A real plan returns it.

Actionable takeaway

Write a one-page "Concentrated Stock Term Sheet" for yourself before you place any hedge:

- *Position facts:* number of shares, approximate cost basis, approximate unrealized gain, and what percentage of your net worth it represents.
- *Constraints:* blackout windows, insider status, Rule 144 considerations (if applicable), employer policy on derivatives, and how fast you can realistically sell.
- *Non-negotiable objective:* the maximum loss you are willing to tolerate over the next 12 months without changing your life plan.
- *Decision rule:* whether you will (a) sell a fixed amount on a schedule, (b) collar a defined number of shares, or (c) do both.

Then take that sheet to two places: your company compliance contact (to confirm what's permitted) and a tax professional (to sanity-check constructive sale and related tax risks). Concentrated stock is one of the few situations in investing where a single "small" mistake can cost more than years of careful saving. A couple of conversations upfront can buy you something priceless: protection that actually sticks.

Chapter 8

Behavior and Governance

You can build the mathematically perfect hedge, but it becomes worthless the moment you panic and sell it. The greatest threat to your downside protection isn't market volatility, rising interest rates, or black swan events-it is the face staring back at you in the mirror. In this final chapter, we move away from Greeks and derivatives to tackle the psychology of insurance. We will explore why our brains are wired to hate paying for protection when the sun is shining, and how to build a 'contract with yourself' that ensures your safety net stays in place when you need it most.

8.1 The Enemy in the Mirror: The Psychology of Protection

The cruelest thing about hedging is that it works *best* when it feels *worst*. In a long, sunny market, protection looks like a leak in the hull: small, persistent, and apparently self-inflicted. Your portfolio is up, your friends are up, headlines are upbeat, and your hedge statement shows a steady drip of premiums paid for disasters that never arrive. The mind does what the mind always does: it tries to *stop the pain* it can see.

That's the first twist to internalize: the pain of hedging is *certain and frequent* (you pay premiums, month after month), while the benefit is *uncertain and rare* (it shows up in storms, not in clear skies). Human brains are not built to love that trade. They're built to hate it, rationalize it, and eventually sabotage it.

Two biases do most of the sabotage work: *recency bias* and *regret*. Recency bias is the mind's habit of overweighting whatever just happened and underweighting what happened before. Regret is the emotional tax we pay when we feel we made a "stupid" choice, especially one that shows up as a line item loss. Together they form a surprisingly reliable machine that convinces otherwise intelligent investors to drop protection at the exact wrong time, then buy it back at the exact wrong price.

The emotional math of ``wasted" premiums

Picture a simple, familiar moment. You buy protective puts (options that gain value when the market falls) or you run a collar, or you allocate a small slice to a strategy meant to spike during crashes. The market rises anyway. Your hedge loses money. You do it again. The market rises again. Your hedge loses money again.

At first, you shrug. You even feel responsible. "This is what I signed up for," you tell yourself. But after enough months, a different story starts to form: "I keep *paying* and I keep getting *nothing*."

That story has teeth because it's not just about returns; it's about *identity*. Losing money on a hedge feels like being wrong in slow motion. And because the loss is explicit—a premium, a drag, a negative P&L line—your brain tags it as a mistake that needs to be corrected. Meanwhile, the benefit (the catastrophe that didn't happen) is invisible. There's no statement line that says: "You did not lose 38% this quarter."

This is where regret becomes an engine. Regret doesn't ask, "Was my decision sensible given the range of possible futures?" Regret asks, "Did the future embarrass me?" If the market didn't crash, the hedge feels embarrassing. If the market did crash and you weren't hedged, the lack of hedge feels embarrassing. Either way, the mind wants to avoid the emotion "next time". The trouble is that "next time" is guided by the freshest feelings, not the full distribution of outcomes.

Recency bias: the tyranny of the latest chart

Recency bias is the reason a calm market feels like a permanent climate instead of a temporary season. After a few quiet years, the last 36 months start to masquerade as a law of nature: volatility is "over," dips are "buyable," central banks are "supportive," innovation is "unstoppable," or whatever story happens to be fashionable.

In that mental environment, hedging begins to look like superstition. The logic becomes seductively practical:

- "Why pay for protection when dips keep bouncing?"
- "I can always buy a hedge *if* things get ugly."
- "My hedge is a drag; I'm falling behind."

Each sentence sounds reasonable in isolation. Together they create a trap: you train yourself to hedge only when the recent past has been scary. Unfortunately, scary recent past and cheap protection almost never coexist.

Markets tend to be most expensive to insure when the demand for insurance is highest. Volatility jumps, option prices jump, and the premium you pay is no longer just "maintenance." It's panic pricing. Recency bias doesn't merely nudge your timing a little late; it can flip your entire strategy from *systematic* to *reactive*.

Insurance fatigue: when boredom cancels the seatbelt

Insurance fatigue is what happens when you do the right thing for a long time and get "punished" for it by normal market behavior.

A hedge that protects against crashes will often lose money in ordinary markets, because ordinary markets are not crashes. If you buy put options and the market grinds higher, the options decay (options lose value as time passes if the market doesn't move in the direction you need). If you hold a tail-risk fund, you may endure long stretches where it bleeds while stocks trend up. If you run a collar, you may give up some upside in exchange for downside protection, and the upside you gave up will feel, emotionally, like money you *should have had*.

The fatigue arrives quietly. It's not the dramatic moment of capitulation; it's the slow, justified decision:

> "I'll stop for a bit. Volatility is low. I'll restart later."

This is how portfolios lose their fire extinguisher: not because the owner hates safety, but because the owner gets tired of paying for a thing that, in good times, looks unused.

Right about here, the market often hands you a final piece of psychological bait: the hedge expires worthless *again*. The mind seizes on the pattern. "See? It never pays." That's recency bias turning a streak into a belief.

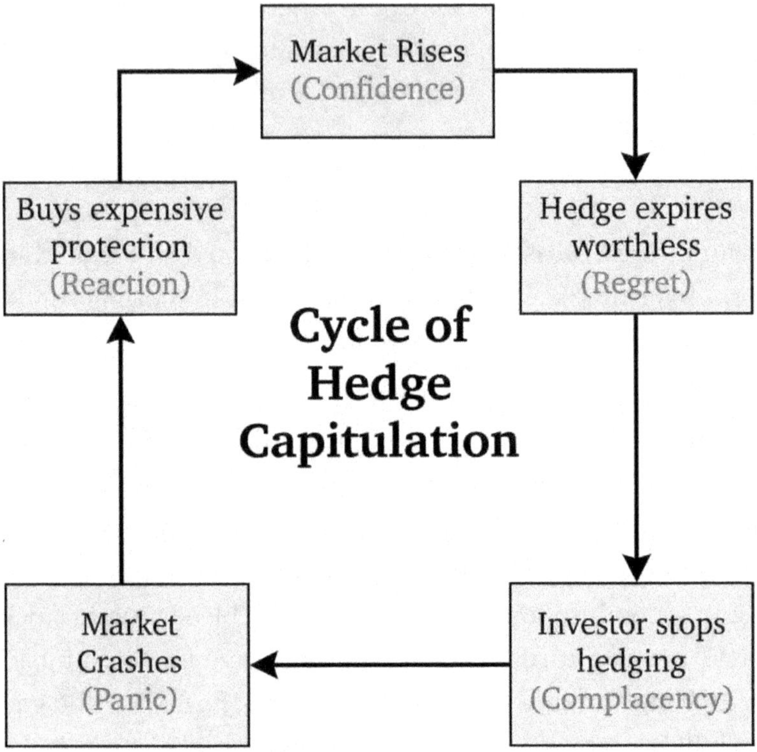

Regret has two faces: ``I paid for nothing" and ``I didn't pay, and now I'm crushed"

Regret is a shapeshifter. In bull markets, it wears the face of the *premium*: "I'm throwing money away." In bear markets, it wears the face of the *missed hedge*: "I was irresponsible."

The reason this matters is that regret doesn't just hurt; it changes behavior. A person trying to avoid regret starts optimizing for emotional outcomes instead of portfolio outcomes. That's when you see the classic pattern:

- after a calm period, protection is reduced or canceled (to avoid the regret of "wasted premiums");
- after a sharp selloff, protection is added aggressively (to avoid the regret of being unprotected).

This pattern feels like learning. It feels adaptive: "I'm responding to new information." But it's often just a sophisticated form of chasing—not chasing performance, but chasing emotional relief.

The deeper irony is that the regret-minimizing choice in the moment is frequently the risk-maximizing choice over time. Canceling protection in calm markets reduces visible costs, but increases the chance of a catastrophic, irreversible drawdown. Buying protection in a panic reduces emotional exposure to headlines, but locks in high implied volatility (expensive options) and can turn the hedge into a short-term consolation prize rather than a long-term stabilizer.

The ``sunk cost" trap: treating hedge premiums like a scorecard

A hedge premium is not a failed trade just because it expired worthless. It is the cost of having had protection *during that period*. If your house doesn't burn down this year, the homeowner's insurance premium was not a "loss" in the way a bad stock pick is a loss. It was rent paid to live in a world where one accident doesn't ruin you.

But portfolios come with a special kind of accounting cruelty: the cost of protection is measurable and frequent, while the avoided disaster is hypothetical. That makes it easy to commit the sunk cost fallacy in reverse. Instead of saying, "I've already paid so I should keep paying," investors often say, "I've already paid so much and got nothing—I should stop." The past premiums become a grievance.

There's a more useful frame that respects both the math and the psychology:

> *A hedge is not something you win on. It's something you maintain.*

Maintenance is boring by design. You don't celebrate that your brakes worked. You don't brag that your roof didn't leak. You just keep going because the alternative is a sudden, expensive failure.

Once you adopt the maintenance frame, the right question changes from "Did my hedge make money?" to "Did my hedge keep my portfolio within a range I can live with?" That's a different scoreboard. And it's a scoreboard that doesn't collapse under regret.

Myopic loss aversion: when you check too often, everything feels broken

There's another psychological accelerant that turns minor hedge losses into major frustration: checking too often.

If you look at a hedge day-to-day or week-to-week, it will frequently look like a mistake. Protective structures often have negative carry (a cost to hold). Their job is not to look smart in quiet weeks; their job is to respond in ugly ones. When the evaluation window is too short, the hedge appears "wrong" most of the time.

This is myopic loss aversion in plain clothes: loss aversion is the fact that losses hurt more than gains feel good; myopia is the fact that frequent observation makes the losses feel constant. Together, they create a behavioral recipe for abandoning risk management. You're not only paying the premium; you're paying the emotional premium of repeatedly watching it bleed.

A practical implication follows: the cadence at which you review a hedge should match the hedge's purpose. If the hedge is meant to protect against multi-month drawdowns or crisis regimes, judging it every few days is like judging a smoke alarm by how often it rings during dinner.

Why ``I'll hedge when I need it'' is usually a fairy tale

The idea sounds comforting: keep the upside, then buy protection when trouble appears. In real time, trouble arrives with two complications:

1. *Your brain won't agree on what counts as trouble.* A 3% dip feels like noise in a bull market and like the start of the end in a fragile one. Recency bias rewrites your thresholds.
2. *The market reprices protection quickly.* When fear hits, option prices tend to rise. You can still buy protection, but it's often not the bargain you imagined in the calm.

The second point is mechanical, but the first is psychological and more dangerous. Without rules, "I'll hedge when I need it" becomes "I'll hedge when I *feel* like I need it." And feelings are notorious market timers.

This is how investors end up owning the most protection at the moment it offers the least expected value: after a volatility spike, after headlines have already landed, after the first leg down has already done its damage. It's not because they're irrational; it's because they're human—and humans prefer buying relief to buying discipline.

Regret avoidance and the weird urge to *undo* a hedge

One of the stranger behaviors in hedging isn't just dropping protection; it's trying to "make back" hedge costs.

An investor pays premiums for a year, feels annoyed, then decides to sell extra calls, increase leverage, chase higher yield, or concentrate into "safer" stocks to compensate for the perceived drag. This is regret management masquerading as optimization. The mind tries to erase the emotional stain of "I paid for nothing" by taking new risks that feel productive.

The portfolio doesn't experience this as a neat accounting fix; it experiences it as a governance breakdown. You didn't merely cancel insurance. You added hidden insurance risk elsewhere. It's like canceling health insurance and then deciding the solution is to drive faster because you "saved money."

If you recognize this impulse in yourself, that recognition is valuable. The urge to *undo* hedge costs is a signal that your brain is treating the hedge like a bet you're trying to win, rather than a constraint you're choosing to live within.

A simple way to catch recency bias in the act

Recency bias thrives on vague language: "things feel different," "the market is strong," "volatility is dead," "this dip is nothing," "this time is scary."

A good antidote is to force yourself into numbers and time horizons:

- "If a 25% drawdown happened over the next 12 months, what would I do?"
- "If volatility stayed low for another 24 months, would I keep paying?"
- "If my hedge cost 1% per year for five years and then saved me 15% in a crash, would I still want it?"

Those questions are not forecasts. They're psychological stress tests. They push your brain away from the last three months and back toward the full range of possible markets. Recency bias hates ranges. It loves narratives.

Turning the emotional cycle into a deliberate choice

It's tempting to treat these biases as personal flaws. They're not. They're default settings.

The problem isn't that you feel regret when a hedge expires worthless. The problem is that regret quietly hijacks the control panel. The problem isn't that calm markets make you complacent. The problem is that complacency changes your rules without announcing itself.

A well-designed hedging program does two things at once:

1. It acknowledges that protection costs money and will often look “wrong” in good times.
2. It reduces the number of emotional decisions required to keep protection in place.

If your hedge requires you to re-justify it every month, you're asking a tired, dopamine-seeking brain to repeatedly choose vegetables over dessert. Eventually, dessert wins.

So the goal becomes less “be more disciplined” and more “need less discipline.”

Actionable takeaway: write a one-page ``anti-regret" contract with yourself

Do this today, in plain language, and keep it where you can actually find it (notes app, printed page, whatever you'll see during stress):

> *For the next 12 months, I will treat hedging costs as maintenance, not as performance. I will not cancel protection because of recent calm markets, and I will not increase protection because of recent panic headlines. I will only change my hedge if (1) my target hedge budget changes, or (2) my portfolio risk changes in a way I can describe in one sentence. I will review the hedge on a fixed schedule: ___ (monthly/quarterly), not daily.*

Then add two numbers beneath it:

- *Annual hedge budget (as % of portfolio):* ____
- *Maximum drawdown I refuse to tolerate without protection:* ____

Those blanks are not paperwork. They're your defense against recency bias and regret—a way to keep the enemy in the mirror from quietly rewriting the plan when the market starts whispering in your ear.

8.2 Tying Yourself to the Mast: The Personal Policy Statement

The market doesn't defeat most hedgers with a single dramatic punch. It wins by wearing them down until they start improvising. First, the hedge "just this month" gets skipped. Then the sizing becomes a vibe instead of a number. Then a scary headline arrives and suddenly the portfolio is being managed by adrenaline.

A Personal Policy Statement is how you stop negotiating with yourself at the worst possible time.

Think of it as a written set of *if-then* promises you make while you're calm, caffeinated, and capable of thinking in full sentences. The point isn't to become rigid. The point is to become consistent. If your hedge is meant to be portfolio infrastructure, it needs a maintenance manual—not a mood ring.

Why writing it down changes everything

There's a strange magic to ink. A plan that lives in your head is easy to "edit" without noticing. A plan that lives on paper resists revision because it makes your future self accountable to your past self. That accountability is the whole game.

Most investors already have an informal policy statement, even if they've never written one:

> "I hedge when it feels scary, and I stop hedging when it feels dumb."

It's honest. It's also a blueprint for buying protection high and canceling it low.

A written policy flips the burden of proof. Instead of needing to justify *maintaining* a hedge every month, you have to justify *breaking* the rule. That one switch is often the difference between a hedge that exists when it's needed and a hedge that exists only when it's expensive.

The goal: fewer decisions, not better predictions

A good policy statement is not a forecast. It doesn't say, "A crash is coming." It says, "If a crash comes, I will already have decided what to do."

That's a subtle but powerful shift. Forecasts ask you to be right about the market. Policies ask you to be right about *your own behavior*. One of those jobs is nearly impossible. The other is achievable with a page of rules and the humility to follow them.

If this feels like lowering ambition, good. Hedging isn't about heroic intelligence. It's about preventing one bad season from changing the rest of your financial life.

Personal Policy Statement vs. professional IPS (and why you should steal the professional parts)

Institutions don't manage money with vibes. They use an Investment Policy Statement (IPS): a document that defines objectives, risk tolerance, constraints, and the rules for implementing and reviewing the portfolio. Individuals can borrow that structure and make it personal, practical, and short enough to actually use.

A Personal Policy Statement for hedging answers five questions:

1. *What is the hedge supposed to do?*
2. *How much am I willing to pay for it?*
3. *When do I add, reduce, or take profits?*
4. *Who is allowed to change these rules?*
5. *How often do I review results without panicking?*

If any of those are fuzzy, your future self will "fill in the blanks" during stress, which is a polite way of saying you'll improvise.

Step 1: Write a one-sentence purpose that a tired person can understand

If the purpose statement is vague, the hedge will be vague. Vague hedges become expensive experiments.

Strong purpose statements are concrete and tied to lived experience:

- *"Limit portfolio drawdowns to about 15–20% in a fast equity selloff."*
- *"Reduce the chance that I sell equities in panic during a large decline."*
- *"Create a source of liquidity I can rebalance from when stocks fall hard."*

Notice what's missing: "beat the market," "make money in crashes," "time volatility." Those can happen, but they make terrible north stars. A hedge that tries to be a profit center often turns into a strategy you abandon the moment it stops being exciting.

A practical test: if your purpose statement can't survive a normal, boring year—a year where nothing blows up—it's not a purpose statement. It's a prediction.

Step 2: Set a hedge budget that your ego can't quietly overrule

You need a budget for protection the same way you need a budget for home repairs. Not because you enjoy the expense, but because surprises are worse.

A hedge budget is typically expressed as a percent of portfolio value per year. Call it your *carry allowance*: the amount of "drag" you're willing to tolerate in exchange for sleeping through storms.

Here's the part people skip: the budget should be tied to your behavior, not your spreadsheets.

If your backtest says you can tolerate a 2% annual hedge cost but you find yourself getting angry at 0.7%, the backtest is irrelevant. The only sustainable budget is the one you can keep paying in a bull market without feeling like a fool.

A simple way to choose a starting point is to answer two questions honestly:

- "If this hedge costs $X this year and nothing bad happens, will I resent it?"
- "If I *don't* hedge and a bad year hits, will I do something self-destructive (sell, lever down at the bottom, abandon the plan)?"

Your budget should be high enough to matter in a drawdown, and low enough that you can keep it on the books when markets are calm.

Step 3: Define what you are hedging (you can't insure a fog bank)

Hedging "the portfolio" sounds tidy until you remember what most portfolios actually are: a mix of broad index funds, a few beloved stocks, maybe some credit exposure, some international, some cash, some private investments, and a handful of positions you pretend are "long term" but check twice a day.

So specify the exposure:

- "I hedge my U.S. equity allocation (S&P 500 / total market proxy)."
- "I hedge *only* the portion I'd be tempted to sell in a crash."

- "I do not hedge bonds/cash; they are part of the ballast."

This is not pedantry. It prevents a common failure mode: buying a hedge on one index while taking concentrated risk somewhere else, then feeling betrayed when the hedge doesn't "save" the thing you actually care about.

If you've got concentrated positions, say so. Concentration is where hedging dreams go to die unless you acknowledge it up front.

Step 4: Choose triggers that are boring on purpose

Most trigger ideas start romantic and end tragic. The romantic version: "I'll add protection when the market looks overvalued." The tragic version: you spend years arguing with yourself about what "overvalued" means.

Good triggers are dull, observable, and hard to litigate.

Three trigger styles work well for individuals:

1) Time-based maintenance (the default that wins by being unsexy).

This means you hedge continuously and refresh on a schedule:

> "On the first trading day of each month, I roll my protection to maintain a constant horizon."

You'll never nail the perfect entry. That's the point. You're buying process, not timing.

2) Drawdown-based actions (because pain is measurable).

Drawdowns are what investors actually experience. So you can write triggers like:

> "If the portfolio falls 5% from its high-water mark, I will *not* cancel hedges. If it falls 10%, I will rebalance and evaluate monetizing part of the hedge."

This replaces "panic decisions" with "pre-approved responses."

3) Volatility regime triggers (useful, but only if you keep them simple).

Volatility indices like the VIX can act as a rough fear gauge. You can specify:

> "If VIX is above 30, I do not initiate new long-vol positions unless they are part of my scheduled roll."

That single sentence can save you from the classic mistake of buying expensive protection in the middle of a fire.

You can combine triggers, but resist the urge to build a cockpit. Complex trigger systems tend to fail in exactly the way you'd expect: they give you more levers to pull when you're emotional.

Step 5: Pre-write the actions (otherwise your trigger is just an alarm)

A trigger without an action is a smoke detector with no exit plan.

Actions should be specific enough that you could hand the page to someone else and they'd know what to do. Examples:

- "If drawdown > 10%, take profits on 20% of hedge gains and move proceeds to cash for rebalancing."

- "If drawdown recovers to within 3% of prior peak, rebuild hedges back to target notional over the next two roll dates."
- "If the hedge cost exceeds my annual budget mid-year, reduce size at the next roll rather than canceling immediately."

Notice the hidden psychological benefit: these actions create *permission*. In a crash, you won't be wondering whether it's "okay" to take profits on protection or "too early" to rebalance. The permission is already granted by the calm version of you.

Step 6: Assign decision rights (yes, even if it's just you)

This sounds corporate until you live through a real selloff and realize you have multiple selves:

- the calm self who built the plan,
- the scared self who wants to tear it up,
- the bored self who wants to optimize it into oblivion.

Decision rights are how you keep the wrong self from taking over. Write something like:

> "I can change hedge rules only during scheduled reviews. Outside of that window, I can execute prewritten actions but not redesign the program."

If you have a spouse/partner involved in finances, this is even more powerful:

> "Any decision to cancel protection requires a 48-hour waiting period and a second signature."

That's not a lack of trust. It's a speed bump for panic.

This is the same behavioral logic behind commitment devices: people are often happy to precommit to good behavior when they are not under pressure, because they know a pressured version of themselves will be less wise.

Step 7: Set a review cadence that prevents both neglect and obsession

A hedging program dies in two ways:

- *neglect*: you forget to maintain it and discover the "protection" expired months ago;
- *obsession*: you watch it daily, hate how it behaves, and start tinkering.

Write a review cadence that matches the hedge's intent. A common approach:

- *maintenance check*: monthly (did I roll, did sizing drift, did anything break?),
- *strategy review*: quarterly or annually (is the program still aligned with my life and risk tolerance?).

Add one more line that feels almost too obvious:

> "I do not judge hedge success by whether the hedge itself made money."

That sentence is a fence around your attention. Without it, you'll naturally grade the hedge like a stock pick, and you'll drift back toward regret-driven decisions.

A practical template you can steal (keep it to one page)

Below is a compact format that fits on a page and covers the bases. It's not meant to be pretty; it's meant to be usable.

Personal Policy Statement: Portfolio Protection	
Purpose	(One sentence.) Example: "Cap portfolio drawdown near 15–20% during fast equity selloffs so I can rebalance instead of panic-selling."
Hedged exposure	Example: "U.S. equity index funds (approx. 60% of portfolio). Concentrated single stocks are not assumed hedged unless explicitly stated."
Hedge budget	Example: "Up to 1.0% of portfolio value per year in expected carry cost. If exceeded, resize at next roll, do not cancel midstream."
Maintenance rule	Example: "Maintain continuous protection with a 3-month horizon; roll monthly on the first trading day."
Triggers	Example: "If portfolio drawdown > 10%, initiate rebalance plan and evaluate taking partial hedge profits. If VIX > 30, do not add discretionary protection."
Actions	Example: "At > 10% drawdown: monetize 20% of hedge gains into cash; rebalance 1/3 of the equity shortfall from cash/hedge profits. Repeat at > 15% drawdown."
Decision rights	Example: "Rules may be changed only during quarterly review. Emergency changes require 48-hour waiting period."
Review cadence	Example: "Monthly maintenance check; quarterly alignment review; annual deep review for budget/purpose."

If you can't fill a row, that's not a failure. That's a flashlight. It shows exactly where your future self would have been forced to improvise.

Common failure points (and the small edits that fix them)

Failure point: the policy is too ambitious.

If your hedge budget is set by what looks good in a spreadsheet rather than what you can emotionally sustain, you'll eventually quit. Fix it by lowering the budget and focusing on consistency. A small hedge you keep is usually better than a big hedge you abandon.

Failure point: the triggers are too clever.

If your triggers require interpretation ("overvalued," "weak breadth," "macro risk rising"), you'll spend your stress debating definitions. Fix it by using measurable triggers (time, drawdown, simple volatility regime) and placing the clever ideas in the "annual review" bucket.

Failure point: you didn't define what "stop" means.

Some people write policies to add protection, but not to reduce it. Then they cling to hedges long after the reason has passed, because selling a hedge feels like jinxing safety. Fix it by prewriting a profit-taking rule and a rebuild rule. Symmetry prevents superstition.

Actionable takeaway: draft the ``stress-proof" version in 30 minutes

Set a timer for 30 minutes and write a one-page Personal Policy Statement using the table above. Two rules make it work:

1. Use plain words your future self won't misread under stress.

2. Wherever you feel tempted to write a prediction, replace it with a trigger and an action.

Then do the final, crucial step: sign it and date it. Not because it's legally binding, but because it turns the page into a commitment device. When the market starts arguing with you, you'll have a calm voice on paper that already won the argument.

8.3 The Scoreboard That Matters: Evaluating Effectiveness

A hedge has an image problem. When it does its job, it often looks like it's *losing*.

You pay option premiums. You cap upside with a collar. You hold something "boring" that doesn't join the party when stocks sprint. Month after month, the hedge prints small red numbers, and your brain treats those numbers the way it treats a bad restaurant bill: proof you were taken for a ride.

That's the wrong scoreboard.

The right scoreboard isn't the hedge's P&L in isolation. It's what happened to *your total portfolio* on the days when the market stopped being polite. If you judge a seatbelt by how stylish it looked on your commute, you'll eventually stop wearing one. If you judge it by what happened in the crash, you'll never drive without it again.

The most common mistake: asking a hedge to be a profit center

People love the idea of a hedge that "pays for itself." In practice, that desire quietly changes the hedge's DNA. Instead of being a stabilizer, it becomes a side hustle. Instead of being sized to meaningfully reduce drawdowns, it's sized to look clever. Instead of being maintained systematically, it's traded tactically.

The danger isn't that you try to make money. The danger is that you start measuring your hedge like a stock pick:

> "Is this trade up or down?"

That question is addictive because it's immediate. It also misses the point. A hedge can lose money for years and still be one of the best decisions in the portfolio—because it changes what you do (and what you don't do) when markets are ugly.

A hedge is closer to a circuit breaker than a lottery ticket. If you insist it should constantly generate returns, you'll end up with something that behaves like a speculative position—and then it will fail at the one job you hired it for.

A better question: did the hedge change my worst month?

Most long-term wealth destruction doesn't happen in average months. It happens in a handful of awful ones, clustered together, when the market feels like it's trying to teach you a lesson.

The hedger's job is to improve the portfolio's experience in that window:

- smaller peak-to-trough drawdown,
- fewer forced sales,
- more ability to rebalance into cheap risk,
- less chance you abandon the plan entirely.

None of those show up cleanly in "hedge return." They show up in the portfolio's path.

And the path is the thing that matters, because portfolios don't compound on average. They compound through time, and time includes cliffs.

The danger of the clean metric: volatility can lie

Investors love metrics that look precise. Volatility (the standard deviation of returns) is one of those metrics: it gives you a single number, a neat summary of wiggles. But volatility can be the wrong scoreboard for protection.

Why? Because volatility treats upside and downside as equally "bad." A portfolio that rockets up and then drifts has high volatility. A portfolio that grinds down in a slow bleed can have lower volatility. Your stomach knows which one is worse, but the metric doesn't.

Downside protection is usually about *asymmetry*: you're trying to reduce the damage from large negative moves more than you're trying to smooth every normal fluctuation. That means you need tail-aware measures—metrics that care about the left side of the distribution, where the real monsters live.

Metrics that match the actual job

A practical way to evaluate a hedge is to decide, in advance, what "effective" means. Not in poetry. In numbers.

Here are portfolio-level metrics that tend to map to real-world protection goals:

Maximum drawdown (MDD). This is the deepest peak-to-trough decline over a period. It's brutally honest. It answers: "How bad did it get before it got better?" If your hedge meaningfully reduces MDD during crisis windows, it's doing important work even if it bleeds in calm years.

Drawdown duration. Two portfolios can have the same maximum drawdown but very different lived experiences. One falls 25% and recovers in four months. Another falls 25% and takes three years to recover. The second one is where panic-selling is born. A hedge that shortens drawdown duration may be worth its cost even if it doesn't create spectacular hedge P&L.

Worst-month (or worst-quarter) return. This is a simple, visceral metric. If the unhedged portfolio's worst month is −12% and the hedged portfolio's worst month is −8%, that delta can be the difference between "unpleasant" and "unbearable," especially for retirees or anyone making contributions and withdrawals.

Left-tail risk metrics (VaR, CVaR/Expected Shortfall). Value at Risk (VaR) is a threshold: "How bad might it get with X% confidence over a period?" Expected Shortfall (also called Conditional VaR or CVaR) asks a more honest question: "If we're in the worst X% of outcomes, how bad is it *on average*?"

For protection, Expected Shortfall often tells the better story, because it focuses on the damage in the tail rather than a single cutoff.

Lower Partial Moments (LPM). This sounds academic until you realize it's just "how often and how severely do returns fall below a threshold I care about?" If your threshold is 0% (don't lose money) or −5% (don't get punched too hard in a month), LPM translates hedging success into plain English: fewer and smaller bad outcomes.

These are not trophies to collect. Pick one or two that match your purpose statement and stick with them. The biggest evaluation error isn't choosing an imperfect metric; it's changing metrics every time the hedge feels emotionally inconvenient.

The ``fire extinguisher problem" and the missing counterfactual

Hedge evaluation is hard because the most important result is a counterfactual: what would have happened *without* the hedge?

If the market doesn't crash, the hedge looks unnecessary. If the market crashes, the hedge looks brilliant. In both cases, the mind tries to rewrite history:

- "I shouldn't have paid for all that protection."
- "I should have bought more protection."

Both statements can be true in hindsight and useless in practice.

A better approach is to build your own counterfactual in a controlled way. Maintain a simple "shadow" benchmark: what your portfolio would have done if you did nothing but hold your strategic allocation. Then compare your actual portfolio to that benchmark during stress windows and over full cycles.

This is also where public benchmarks can help. Systematic protection indices (for example, put-protection or collar benchmarks) provide a reality check: protection tends to reduce drawdowns and severe down months, but it typically gives up some upside. If your hedge results look wildly better than those benchmarks without a clear reason, you may be underestimating risk or overfitting a story to a lucky period. If your results look wildly worse, you may be paying too much, sizing poorly, or hedging the wrong exposure.

The hidden win: protection can improve compounding even if it drags on returns

There's an argument that makes long-term investors sit up straighter: trimming the left tail can improve long-run outcomes even when the hedge has a negative expected carry.

Compounding is path dependent. A 50% drawdown requires a 100% gain to get back to even. That math is not negotiable. If a hedge reduces the depth of the crater, it can reduce the required recovery and keep the portfolio on a smoother compounding track.

This is why some institutional debates about tail hedging sound paradoxical: "It may not raise average returns, but it can still be valuable." That's not wordplay. It's recognizing that avoiding catastrophic paths can matter more than squeezing a few extra basis points out of the average year.

A practical annual review that doesn't turn into a witch trial

Most hedge reviews go off the rails because they start with an accusation:

> "Why did we lose money on this?"

That question invites defensiveness and tinkering. A better review starts with alignment:

> "Did this program do what it was designed to do,
> at a cost I agreed to pay?"

A clean annual review can be done on one page:

Review Item	*What to record*
Purpose alignment	Does the hedge still match my goals and time horizon? Any life changes (retirement date, liquidity needs) that alter the purpose?
Budget adherence	Hedge carry cost vs. budget (in % of portfolio). If over budget, why (volatility regime, sizing drift, poor execution)?
Crisis-window behavior	In the worst month/quarter of the year, how did the total portfolio perform vs. an unhedged benchmark?
Drawdown metrics	Maximum drawdown and drawdown duration for hedged vs. unhedged benchmark over the review period.
Governance integrity	Did I follow my own maintenance and trigger rules? If not, what broke: process, tools, or emotions?
Implementation notes	Any slippage from spreads/fees? Any concentration exposures not covered? Any operational failures (missed rolls, wrong notional)?

The point of this review is not to punish the hedge for being a hedge. It's to verify that you're getting the *type* of protection you intended, without silent drift in cost, exposure, or behavior.

The only performance question that really matters

When investors say, "I want a hedge that works," what they often mean is: "I want a hedge that makes me *feel* smart." The market is rarely willing to provide that feeling on schedule.

A more honest question is:

> "If the next major correction starts tomorrow, will I be able to stick to my plan?"

If the hedge increases the odds that you stay invested, rebalance rationally, and avoid permanent damage, it's not just a derivative position. It's behavioral scaffolding.

That scaffolding doesn't show up in the hedge's standalone return. It shows up in the fact that you didn't do something irreversible at the worst possible moment.

The picture that ends the argument

Near the end of a selloff, portfolios start to feel personal. The numbers look like judgment. That's when it helps to see, plainly, what "drag vs. protection" can mean in dollar terms: a deeper crater versus a shallower one, and the space between them—the room where your future compounding still fits.

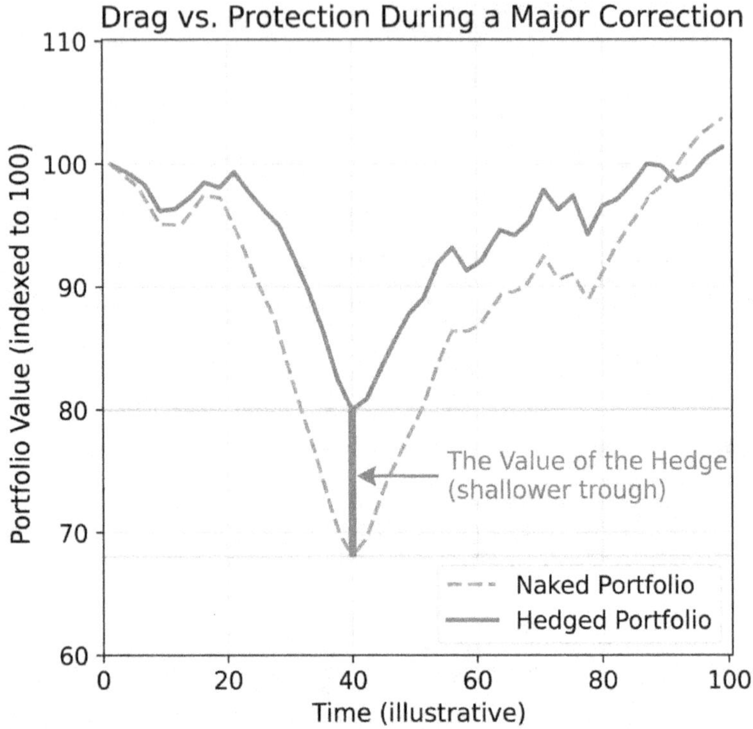

Actionable takeaway: score your hedge like an engineer, not a gambler

Before you change a single knob on your protection program, write down two numbers and one window:

- *Maximum acceptable drawdown:* ____ %
- *Maximum annual hedge budget:* ____ %
- *Stress window for evaluation:* worst month and worst quarter each year

Then, once per year, compare your *total portfolio* to a simple unhedged benchmark on those exact measures. If the hedge reduced the depth or duration of the worst decline within your budget, it did its job—even if the hedge itself "lost money." That's the scoreboard that matters.

www.ingramcontent.com/pod-product-compliance
Lightning Source LLC
LaVergne TN
LVHW052027200926
848481LV00005B/1602
* 9 7 9 8 8 9 6 6 5 3 0 0 4 *